Productivity and Reliability-Based Maintenance Management

Matthew P. Stephens

Upper Saddle River, New Jersey
Columbus, Ohio

Editor in Chief: Stephen Helba
Executive Editor: Debbie Yarnell
Associate Editor: Kimberly Yehle
Production Editor: Louise N. Sette
Production Supervision: Custom Editorial Productions, Inc.
Design Coordinator: Diane Ernsberger
Cover Designer: Keith Van Norman
Production Manager: Brian Fox
Marketing Manager: Jimmy Stephens

This book was set in Palatino by Custom Editorial Productions, Inc. It was printed and bound by R. R. Donnelly & Sons Company. The cover was printed by Phoenix Color Corp.

Pearson Education Ltd.
Pearson Education Singapore Pte. Ltd.
Pearson Education Canada, Ltd.
Pearson Education—Japan

Pearson Education Australia Pty. Limited
Pearson Education North Asia Ltd.
Pearson Educación de Mexico, S.A. de C.V.
Pearson Education Malaysia Pte. Ltd.

10 9 8 7 6 5 4 3 2 1
ISBN: 0-13-096657-6

ad majorem Dei gloriam

ABOUT THE AUTHOR

Matthew P. Stephens is a professor and a University Faculty Scholar in the Department of Industrial Technology at Purdue University, where he teaches graduate and undergraduate courses in total productive maintenance management (TPM), facilities planning, and statistical quality control. Professor Stephens holds undergraduate and graduate degrees from Southern Illinois University and the University of Arkansas, with specialization in operations management and statistics.

Prior to joining academe, Dr. Stephens spent nine years with several manufacturing and business enterprises, including flatbed trailer and washer and dryer manufacturers. He also has been extensively involved as a consultant with a number of major manufacturing companies.

Professor Stephens has numerous publications to his credit in the areas of productivity and quality improvements and lean production systems. He is the coauthor of *Manufacturing Facilities Design and Material Handling*, 2nd ed. (Prentice Hall, 2000). He has served various professional organizations including the National Association of Industrial Technology and the American Society for Quality, of which he is a Senior Member and a Certified Quality Engineer.

PREFACE

Productivity and Reliability-Based Maintenance Management is intended to provide a strong yet practical foundation for understanding the concepts and practices of total productive maintenance (TPM) management—a proactive asset and resource management strategy that is based on enhancing equipment reliability and overall enterprise productivity. The book is intended to serve as a fundamental yet comprehensive educational and practical guide for departing from the wait-failure-emergency repair cycle that has plagued too many industries and advancing to a proactive and productive maintenance strategy. It is not intended to be a how-to-fix-it manual but instead emphasizes the concept of a world-class maintenance management philosophy to avoid the failure in the first place.

This book serves to fill an immense void in the instructional needs of technology students at all levels. Universities, junior and community colleges, and technical institutes as well as professional, corporate, and industrial training programs can benefit by incorporating these fundamental concepts in their technical and managerial curricula. It can serve as a powerful educational tool for students as well as for maintenance professionals and managers.

In Chapter 1, various types of maintenance organizations and practices are defined and discussed. Through a realistic cost-and-benefit approach, the book introduces the reader to the basic concepts of productivity through proactive maintenance management practices.

The introductory chapter is followed by a discussion of equipment life expectancy, expected failure rates, and the general concepts of reliability in Chapter 2. Basic statistical models and distributions necessary for understanding reliability issues also are discussed. The author embraces the philosophy that concepts, simple or complex, are best appreciated by the reader when they are presented in a clear and easy-to-understand manner. Therefore, all mathematical presentations are reduced to the simplest form.

Chapters 3 and 4 provide a clear presentation of state-of-the-art yet practical preventive and predictive techniques and applications, followed by an in-depth discussion of nondestructive testing and evaluation techniques as they are applicable to maintenance management strategies in Chapter 5. Practical steps for determining current equipment effectiveness and real-world guidelines and examples for implementation of TPM are presented in Chapters 6 and 7. Strong emphasis is placed on operator and employee education, empowerment, and ownership of equipment and processes.

Other important and relevant failure prevention and productivity improvement topics such as benchmarking; cause and effect, root cause analysis; fault tree; and failure mode and effect analysis (FMEA) are presented in Chapter 7. Clear emphasis is placed on the applicability of these tools in the maintenance arena. Chapter 8 takes a practical look at maintenance project management, and the application of various tools is explained. CMMS (computer maintenance management systems) and its role in modern maintenance management are introduced in Chapter 9.

Relevant and real-world case studies are used in most chapters in the Case in Point features, in order to highlight the importance of sound and proactive maintenance practices. End-of-chapter questions summarize and solidify the concepts that are presented in each chapter.

ACKNOWLEDGMENTS

I would like to express my gratitude to all the people who have guided me, encouraged me, and shared their wisdom along this journey. To them, too numerous to name, I am indebted. I am grateful to all organizations and corporations who have generously provided me with state-of-the-art information and technologies to make this book a valuable tool in the field of TPM. I would like to thank Dr. Edie K. Schmidt and her father, Dr. Donald L. Schmidt, for their expertise and contributions in the chapter about project management. I also would like to thank my friends, Dr. Ted Spickler, for his example in event and causal factors diagram; and Dr. Maximo Ortega, for his example in PM cost analysis. I would also like to acknowledge Dr. Maximo Ortega's and Melissa Woods's contributions in development of the instructors' manual. A special thanks goes to my dear friend and colleague Dr. Niaz Latif for his constant support and encouragement. And, a special thanks to you, Christine, for your love and encouragement and for always being there.

I also would like to acknowledge the reviewers of this text: Jeffrey Hoffman, Northern Michigan University; Vedaraman Sriraman, Southwest Texas State University; Toni Doolen, Oregon State University; and Fred Vondra, Tennessee Technological University.

CONTENTS

CHAPTER 3   PREVENTIVE MAINTENANCE   63

CHAPTER 4   PREDICTIVE MAINTENANCE   83

CHAPTER 5   NONDESTRUCTIVE TESTING AND EVALUATION   133

CHAPTER 6   IMPLEMENTING TPM   149

1

INTRODUCTION

Overview

Objectives

At the completion of the chapter, students should be able to

- Define *maintenance*.
- Develop a basic understanding of the role of maintenance in profitability and productivity.

- Understand the primary and secondary goals of maintenance.
- Identify the three types of maintenance activities.
- Develop a basic understanding of TPM.
- Explain the different ways to organize the maintenance department effectively.

1.1   INTRODUCTION TO MAINTENANCE MANAGEMENT

In today's global economy with fierce competition to attain and maintain the competitive edge in productivity and quality, a key factor often is neglected. The *planning* and *managing* of productive maintenance activities in industrial and manufacturing organizations rarely are given the attention they deserve. Whereas industrial managers and corporate leaders fully realize the importance of investment in the latest technologies for quality and productivity improvements, the maintenance of the equipment and technologies does not seem to enjoy the same level of attention and priority. It is ironic that the increased complexity and automation of plant equipment highlights the need for highly skilled maintenance personnel and specialized planning, training, and development of programs for maintenance activities.

Such planning aims at minimizing downtime and provides for the most efficient and effective use of the facilities and equipment at the lowest cost. The mind-set of equating maintenance activities, especially *planned* preventive and predictive maintenance, with expenditures must be changed. The maintenance department should be considered a *profit center* rather a cost center. Consider the most basic definition of profit:

$$Profit = Income - Expense$$

Any unnecessary expenditures that can be avoided by implementing a sound maintenance program, such as downtime, idle equipment and personnel due to equipment breakdown, missed delivery dates, and subsequent loss of customers, will result in reduction of expense, hence an increase in profit. A utility company that is responsible for supplying power to 13 counties in Florida shuts down one of its power-generating units for one to three weeks per year for planned maintenance work. Furthermore, each unit undergoes an exhaustive inspection and overhaul that takes nearly eight weeks once every three years. The maintenance activities entail the performance of several thousand repair and preventive tasks. These planned shutdowns can cost the company an additional $50,000 to $60,000 per day. However, additional costs associated with an unplanned outage could cost from $250,000 to $500,000 per day. This simple example clearly underscores how planned maintenance activities can and do reduce real costs.

1.1.1   Definition of Maintenance

Maintenance can be defined as all activities necessary to keep a system and all of its components in working order. The objectives of any maintenance program should be to maintain the capability of the system while controlling the cost. The components of the cost can be further defined as follows:

- The cost of maintenance labor and material

- The cost of production loss due to an inadequate and ineffective maintenance program

Any deviation or change in a product or a system from its satisfactory working condition to a condition that is below the acceptable or set operating standards for the system can be defined as *failure*. Although all failures do not necessarily result in catastrophes, most are disruptive, inconvenient, wasteful, expensive, and at the very least annoying. Maintenance programs are aimed at eliminating or reducing the number of these failures and the costs associated with them. Widespread automation and mechanization adds to the complexity and the necessity of prudent maintenance programs. Millions of dollars spent on state-of-the art technologies and equipment, along with elaborate training programs, innovative approaches for improved customer service, quality and productivity improvements, inventory management and control techniques, are quickly nullified by unplanned equipment failures and unexpected plant shutdowns.

The severity of these failures and associated costs can give an invaluable indication of the level and the extent to which a maintenance program is required and can be justified. A certain level of maintenance is almost always needed because of natural wear and tear on equipment, tools, and facilities in general, but the maintenance cost must be in line with the savings resulting from such maintenance programs. Figure 1–1 compares the cost of a preventive maintenance program (fix it before it breaks) with the overall cost of failures and the associated repairs. As expected, the cost of a maintenance program increases at higher levels of prevention. Planned shutdowns; inspection of systems and components; statistical, mechanical, chemical, and other required analyses; and replacement of parts and components *before* they are completely worn out all constitute a substantial cost. As this level of prevention increases, however, the costs associated with failures decrease. The question then becomes, "At what point is the ounce of prevention no longer worth the pound of cure?" The optimum prevention level is the point at which the total costs (the cost associated with prevention maintenance plus the cost of repairing failed equipment) are at the minimum. The maintenance manager must therefore determine the cost and the benefits associated

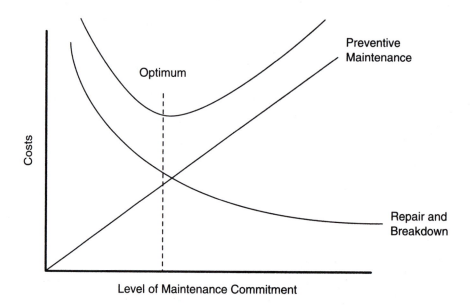

FIGURE 1–1   Comparison of costs at different levels of maintenance

with the unique circumstances of a specific company or department and decide on the appropriate maintenance program.

It is imperative to realize that Figure 1–1 does not represent a universal relationship between the cost of repairs and preventive maintenance for all situations. Although the cost of breakdowns is expected to dramatically decline with the implementation of an effective preventive maintenance program, the curve depicting the increase in the cost of preventive maintenance may assume varying slopes. A variety of factors such as the type and the age of equipment, type of industry, as well as the training and the commitment level of the operating personnel can affect the slope of the curve. Figure 1–2 shows how the slope of the PM cost curve can and does influence the optimal level. Furthermore, we cannot assume that the relationship between the preventive maintenance commitment and the cost always will be linear. Beyond a certain point of maintenance commitment, the curve, and hence the cost, may increase exponentially due to demand for specialized training or special monitoring equipment, as shown in Figure 1–3. Conversely, as an organization matures in its skills to apply and perform preventive and predictive strategies, the cost of such practices may actually decrease over time. Experience with one's own industry and benchmarking against the best in class (discussed in more detail in Chapter 7) are usually good starting points for determining the initial level for preventive maintenance commitment. Once organizations

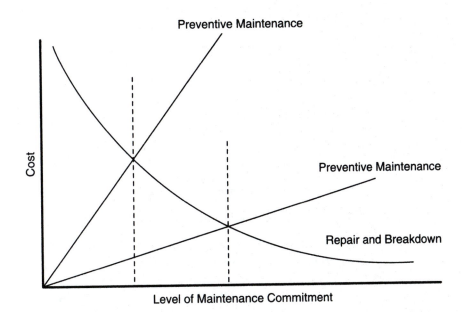

FIGURE 1–2 Maintenance costs comparisons with varying slopes for PM

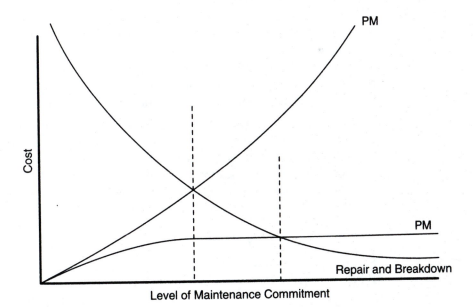

FIGURE 1–3 Maintenance costs comparisons with nonlinear PM

discover the benefits of their preventive maintenance program, the level of commitment gradually increases.

1.2 MAINTENANCE OBJECTIVES

As stated earlier, the overall objectives of a maintenance program should be to maintain the capability of a system while controlling the cost. Within a manufacturing facility, these objectives or goals can be divided into two main categories: *primary* and *secondary goals.*

1.2.1 Primary Goals

Primary goals may include the following:

- Maintaining existing equipment. The timely and appropriate response to equipment failure, reduction of equipment downtime, and an increase in equipment availability can be accomplished by establishing adequate preventive and predictive maintenance programs.
- Equipment inspection, cleaning, and lubrication. This goal may include development of a comprehensive program for operators to perform routine tasks to detect problems before they occur and a comprehensive schedule for regular and routine cleaning and lubrication and equipment.
- Equipment modification, alteration, and installation. These are nonroutine activities and therefore can be schedule during non-peak or slack periods to increase the efficiency and utilization of the maintenance personnel.
- Utility generation, distribution, and management. This covers maintenance and efficient operation of steam, electricity, and so on.
- Maintaining existing building and grounds. This may include building repairs, painting, and a variety of other similar tasks.
- Building modification and alteration. Plant expansions and changes in processes may necessitate other related changes. These tasks, as with equipment installation and modification, often can be scheduled as filler work for slack periods.

1.2.2 Secondary Goals

Secondary functions of the maintenance department may include the following:

- Plant protection and security
- Salvage of obsolete equipment and waste disposal

- Pollution and noise control
- ADA, EPA, OSHA, and other regulatory compliance
- Any other function that may be deemed appropriate by the plant manager

1.3 MANAGEMENT AND STRUCTURE OF THE MAINTENANCE FUNCTION

The nature and the size of the industry determines the structure and the organization of maintenance department; therefore, no single model can serve as the definitive maintenance organization. The processes handed in a plant, whether they are primarily chemical, manufacturing, or other types of activities, determine the need for specific skilled technicians. The size of the facilities and the number of production employees, as well as the degree of automation, can influence the size of the maintenance crew.

Regardless of the organizational structure, maintenance management must include two important functions: *planning* and *scheduling*.

1.3.1 Planning

Planning the maintenance activities is necessary to set goals and objectives and to establish the procedures for accomplishing these goals. An important aspect of planning is setting priority for various tasks, estimating the amount of time required for completing each task, and determining the type of equipment, tools, and labor needed to accomplish the task. The planning process must address such issues as the amount of maintenance needed and the size and the skills of the maintenance crew needed to achieve the maintenance objectives. Planning also is concerned with upgrading and updating the skills of the maintenance personnel, providing for training, and involving the production employees in routine and basic maintenance tasks such a routine cleaning, inspection, and lubrication. Plans must specify which, if any, maintenance functions will be performed by outside contractors and clearly identify these agents.

Planning also establishes limits and tolerances for deviations from set goals. Therefore, a system should be developed to measure the performance of the maintenance department and to compare the actual performance against the objectives and determine the causes of deviation. Based on the feedback, plans should be devised for corrective actions and procedures to implement these corrective actions.

1.3.2 Scheduling

Scheduling is the other important function of a successful maintenance program. Scheduling involves the actual execution of the planned (and sometimes unplanned) maintenance activities. The primary concern is setting the sequence of the outstanding work orders. The scheduler must consider priority, availability of maintenance personnel, and material and maintenance equipment availability. In general, a work order should not be scheduled until all resources are available.

Scheduling requires the prudent use of the resources available and a careful management of the maintenance backlog. Successful scheduling requires a keen knowledge of the time requirements (work measurement) for each task and scheduling techniques such as critical path methods (CPM) and analysis, which are covered in more detail in Chapter 8.

Determining and maintaining a desired level of backlog is critical in the efficient operation of the maintenance department. The level of backlog determines the responsiveness of the maintenance department on one hand and the idleness of the maintenance crew on the other, and provides alternatives for the scheduler. Too much backlog, or a backlog that is continuously growing, can significantly slow the response time and increase the equipment downtime. Too little or no backlog can decrease the productivity of the maintenance crew and increase the cost. As a rule of thumb, an average backlog of two to three weeks is recommended. The optimal crew size can be determined based on the desired level of backlog.

Equation: Crew Size

$$\text{Crew size} = \text{(Scheduled labor-hours per week)} / \text{(Backlog} \times \text{Hours per week)}$$

Example: If the schedule requires 1,400 labor-hours of maintenance work during a 40-hour week, and we want to maintain a backlog of three weeks, then:

$$\text{Crew size (number of employees)} = 1400 / (3 \times 40) = 11.67 \text{ or } 12$$

The equation also can be rearranged to determine the backlog as a function of the crew size, available hours, and scheduled number of hours.

Equation: Backlog

$$\text{Backlog} = \text{(Scheduled hours per week)} / \text{(Crew size} \times \text{Hours per week)}$$

Modern maintenance management planning and control incorporates the use on computerized maintenance management systems (CMMS) and employs various statistical analysis and techniques. Computer simulation and waiting-line theory (queuing theory) are used to determine resource allocation requirements, such as staffing, and for effective planning and scheduling of events. The ability to track and control the backlog, as well as creating and tracking equipment history and work orders, all are important factors in effective management of the maintenance function.

1.4   TOTAL PRODUCTIVE MAINTENANCE (TPM)

1.4.1   Definition

The objective of TPM is to provide a continuous and overall improvement in equipment effectiveness through the active involvement and participation of all employees. TPM is not merely a maintenance program; it is an equipment management program. It combines and promotes the concepts of continuous, total quality improvement and employee empowerment. TPM aims to achieve zero breakdowns and zero defects by making the operator a partner in the maintenance and equipment management efforts.

1.4.2   Operator Responsibility

The operator is the key participant in the TPM environment. The maintenance department takes the role of the advisory group by providing training, setting standards, and performing major repairs.

Instead of relying exclusively on the maintenance department to perform all maintenance tasks, TPM utilizes all available resources including the operators, maintenance personnel, engineers, and the vendors to improve and maintain the equipment at its highest level of performance. Autonomous group of operators are empowered and take ownership of the equipment and assume responsibility for basic and routine maintenance activities, which include the following:

- Housekeeping: Proper organization of the work environment and appropriate locations for all tools, materials, and parts.
- Equipment cleaning: The simplest yet most important step in equipment maintenance and improvement, and at the same time the most neglected step. Equipment cleanliness is the first and most critical step in inspecting and detecting any equipment malfunction. The operator can perform this task on a routine basis with a minimal amount of effort.

- Protection from dirt: Covering machinery and equipment after each use, whenever possible, protects it from dirt and the environment.
- Lubrication: The equipment operator can perform most lubrication on a routine schedule.
- Inspection: Routine inspection can detect vibration, loose bolts, or other obvious abnormalities that are probably more evident to the operator who knows the equipment than anyone else.
- Routine adjustments: The operators can perform routine adjustments to keep equipment operating efficiently.

1.4.3   Obstacles to Achieving Full Equipment Effectiveness

TPM programs strive to reduce and eliminate six significant obstacles to achieving full equipment effectiveness:

1. **Equipment failure**. Direct involvement of the operator as the first line of defense can reduce equipment breakdown.
2. **Setup and routine adjustments**. The machine operator can increase equipment utilization by reducing setup times and performing routine adjustments.
3. **Idling and stoppage**. A trained operator can detect and remedy routine causes of abnormalities and slowdowns due to sensors, blockages, and so on.
4. **Reduced speed**. Trained operators will be able to detect deviations between the expected and the actual speed and take appropriate action.
5. **Defects**. Scrap and defects due to out-of-control conditions can be easily recognized and corrected.
6. **Startup problems**. Problems associated with achieving a stable process are identified and rectified.

1.5   TYPES OF MAINTENANCE ACTIVITIES

Maintenance activities can be grouped into three categories: (1) reactive or corrective maintenance, (2) preventive maintenance (PM), and (3) predictive maintenance (PDM).

1.5.1   Reactive or Corrective Maintenance

As implied by the title, reactive or corrective maintenance is the repair work required after the equipment failure has occurred. Seldom, if ever, does an equipment breakdown occur at a convenient or an opportune time. Even if such failures do not create serious damage to other components and/or injury to people, they cause shutdowns, delay production, and necessitate unexpected and unplanned repairs, and therefore are the most expensive and costly type of maintenance activity. The aim of a proactive maintenance management program should be to reduce the need for this type of maintenance through establishment of PM and PDM whenever feasible. Depending on the nature of the failure, scheduling these repairs often constitutes a high priority and will likely interfere with other planned activities. In some cases when material, equipment, or skilled maintenance personnel are not available, the problem takes on an added dimension of cost.

1.5.2   Preventive Maintenance (PM)

Contrary to reactive maintenance, preventive maintenance takes steps to prevent and fix problems before failures occur. These steps may include proper design and installation of equipment; keeping an accurate history of equipment performance and repairs; scheduled routine inspections and performing necessary upkeep and service; and scheduled cleaning, lubrication, and overhaul. Since these activities are *planned*, shutdowns, when necessary, will not cause undue burden on the production activity, and availability of materials and personnel is ascertained. Most operators with a minimal amount of training can perform most PM activities.

1.5.3   Predictive Maintenance (PDM)

Statistical tools and various instruments and tests, such as vibration analysis, chemical analyses of lubricants, thermography, optical tools, and audio gages, are used to *predict* possible equipment failure. Appropriate preventive maintenance steps can be scheduled and performed based on the data gathered and the analyses performed.

Successful PM and PDM programs require proper identification of all equipment and designation of the required level of PM and PDM. A well-defined inspection schedule and the development of appropriate checklists are necessary, along with trained operators and inspectors, to carry out more sophisticated PDM procedures. A committed budget for this type of maintenance is also necessary.

PDM is necessary to develop an efficient and reliable production system. The returns on such investments are direct and significant; they include quality improvements, long and reliable equipment life, improved safety, and better customer service. PDM will result in increased employee morale due to reduced breakdowns and fewer downtimes, less idle time, reduced needs for spare parts, and reduced overall maintenance costs. PM and PDM and will be discussed in greater detail in subsequent chapters.

Tables 1–1 through 1–3 show the status of maintenance programs and activities for various manufacturing and process industries in North America. These tables contrast the actual percentage of each type of maintenance activity for a given industry with that of the ideal level found in world-class organizations. Tables 1–1, 1–2, and 1–3 compare reactive, preventive, and predictive maintenance activities respectively.

Although the level of various maintenance activities substantially varies for any given enterprise, Table 1–1 shows that currently the overall amount of reactive maintenance in each category is substantially higher than for what could be considered a world-class maintenance operation. Consistent with the data in Table 1–1, Tables 1–2 and 1–3 indicate that in each area of the manufacturing industry, the efforts spent in the areas of preventive and predictive maintenance need to dramatically improve for a maintenance organization to achieve a world-class status.

1.6   MAINTENANCE DEPARTMENT ORGANIZATION

The organization of the maintenance department varies depending on the size and the nature of the enterprise. It may have no formal organization and consist of only one or two individuals who report to the production supervisor and perform the required functions on an as-needed basis. In large plants that require a large crew with varied skills and trades, the maintenance organization becomes more formal and may require its own group of supervisors, planners, and schedulers. In general, the maintenance organization may be a centralized or a decentralized system; each method has its own set of advantages and disadvantages. Experts will agree to disagree on the benefits and pitfalls of each type of system.

1.6.1   Centralized Maintenance Department

A large maintenance organization requiring special material and equipment may lend itself to this type of system, in which crew members may be assigned to performed various tasks in any area of the plant but report to the maintenance department head. This arrangement allows for better utilization of human and equipment resources. Crew members can receive better and more specialized training, and since the overall backlog can be

TABLE 1–1

Current Reactive (Corrective) Maintenance Status in North America

Type of industry	Actual	World class
Assembly	69%	13%
Distribution	34%	17%
Manufacturing (Large)	61%	19%
Manufacturing (Small)	53%	18%
Process	50%	15%
Consultants' Opinion	59%	18%
Weighted Average	55%	18%

Source: Courtesy of Edward H. Hartmann, International TPM Institute, Inc.

TABLE 1–2

Current Preventive Maintenance (PM) Status in North America

Type of industry	Actual	World class
Assembly	29%	53%
Distribution	56%	54%
Manufacturing (Large)	29%	51%
Manufacturing (Small)	34%	52%
Process	34%	42%
Consultants' Opinion	25%	44%
Weighted Average	33%	47%

Source: Courtesy of Edward H. Hartmann, International TPM Institute, Inc.

TABLE 1–3

Current Predictive Maintenance (PDM) Status in North America

Type of industry	Actual	World class
Assembly	7%	39%
Distribution	10%	30%
Manufacturing (Large)	12%	30%
Manufacturing (Small)	12%	32%
Process	15%	42%
Consultants' Opinion	15%	38%
Weighted Average	13%	35%

Source: Courtesy of Edward H. Hartmann, International TPM Institute, Inc.

controlled through more effective planning and scheduling, fluctuation in the number of maintenance personnel can be kept to a minimum. This system allows for efficient control of the inventory of special equipment and material. Centralized maintenance organization provides for better accountability for maintenance activities and the maintenance budget.

Disadvantages of centralization   There are a few disadvantages associated with a centralized department. Since the maintenance crew is dispatched from a fixed location, a substantial amount of time is lost traveling among various locations, reducing the productivity of the maintenance crew. In a large organization, scheduling can become cumbersome and the response time may increase to unacceptable levels. Transportation of tools and equipment from location to location increases material handling and may require special material handling equipment. Supervision of the maintenance crew also may pose a problem.

1.6.2   Decentralized Maintenance Organizations

This approach to the maintenance organization allows various areas, units, or departments to have their own maintenance departments. Obviously, this system allows for a faster response to maintenance needs, reduces travel time and material handling problems, and allows better supervision of the maintenance crew. Furthermore, since the same crew work on the same equipment, they can develop a better understanding of the equipment and its eccentricity or special characteristics. The disadvantages of this system can include duplication of personnel, special maintenance equipment, and material requirements across the plant that add to inefficiency and cost.

1.6.3   Combined System

A blend of the two systems works well for many organizations. A centralized system may handle specialized maintenance operations and have responsibility for the special maintenance equipment and material inventory. Major equipment overall, modification, and rebuilds that require long-term planning and scheduling may be performed by the central maintenance staff. Hiring and retaining specialized maintenance engineers and staffs who serve the entire plant can be easily justified. The central maintenance department also may handle specialized predictive maintenance activities for the entire plant.

Smaller, less specialized crews can be assigned to units in order to reduce response time and handle emergencies in a timely fashion. Local crews also may be in charge of basic and routine preventive maintenance operations.

1.7 MAINTENANCE IN SERVICE INDUSTRIES

Our major focus in this text will be on manufacturing and production industries, but we would be quite remiss if we did not consider, albeit briefly and in passing, some aspect of maintenance in service enterprises. Available data and literature point to significant annual loss of life and money because of improper or insufficient maintenance procedures. These *predictable* and *preventable* maintenance-related failures that can result in loss of life or injury, financial and economic losses, or mere inconvenience and loss of good will, may be related to amusement park equipment that malfunctions, the crash of a jetliner, or delays in providing essential services.

Although a number of factors may result in an accident, the focus of this discussion is those incidents that are specifically maintenance related—accidents that could have and therefore should have been prevented. It is imperative to realize that a maintenance program by itself is not a sufficient condition. Complete adherence and thorough follow-up is equally, if not more important, since an unused maintenance program can create a false sense of security. For instance, it is inconceivable to imagine that any company in the airline industry would not have a state-of-the-art maintenance program sanctioned by various government agencies including the Federal Aviation Administration (FAA). Yet consider the number of *maintenance-related* accidents that occur each year.

Table 1–4 presents the number of aviation accidents reported by the National Transportation Safety Board (NTSB) over a recent six-year period. NTSB is the official investigative agency of the U.S. federal government responsible for investigation and determination of the causes of all

TABLE 1–4

U.S. Civil Aviation Accidents Involving Maintenance as a Cause or Factor

Year	Number of accidents	Accidents with fatalities	Injuries					Total people involved
			Fatal	Serious	Minor	None	Total injuries	
1995	150	25	62	49	70	200	181	381
1996	168	32	161	49	60	711	270	981
1997	149	26	47	41	85	487	173	660
1998	139	23	36	39	54	496	129	625
1999*	78	10	13	17	23	92	53	145
2000*	16	0	0	3	10	13	13	26

Source: United States National Transportation Safety Board (NTSB), Washington, DC.

*Only partial data available at the time of this publication.

air traffic accidents. The table displays only U.S. civil (nonmilitary) aviation accidents that involve *maintenance* as a cause or a factor for the period of 1995 through 2000. A cursory examination of the data may indicate an improving trend for the last two years; however, due to the nature and the extent of inquiry into the causes of accidents and the amount of time that is required to finalize such investigations, the data reported for 1999 and 2000 were incomplete at the time of this writing.

The table lists the number of individuals who were either killed or seriously injured in a variety of civil aviation accidents in which maintenance activities were determined to be either the cause or a contributing factor. In other words, proper maintenance programs and/or adherence to these procedures could have prevented these losses. It is significant to point out that neither noncivilian (i.e., military) mishaps nor financial losses are reported in this table.

The following case study highlights the significance of preventive maintenance programs and how such programs can play a role in air traffic safety and can prevent loss of life. Unfortunately, we often do not need to look far beyond our own industries to see how the lack of proper maintenance programs and practices can cause harm or result in destruction of property and productivity with significant and adverse financial consequences.

 ## THE CRASH OF FLIGHT 261

At about 4:30 P.M. on Monday, January 31, 2000, Alaska Airlines Flight 261 plunged into the Pacific Ocean north of Los Angeles, killing everyone aboard. The Boeing McDonnell Douglas MD-80 was carrying 83 passengers and five crew members.

Although the National Transportation Safety Board (NTSB) is still investigating the exact cause of the crash, and the final report has not been published yet, from the onset of the investigation certain aspects have focused on "maintenance organization and procedures" along with "airline industry lubrication practices."

So far, much of the NTSB's investigation of the crash has focused on the apparent failure of the jackscrew mechanism in the tail section of the plane, which is responsible for raising and lowering the front edge of the stabilizer. According to the investigators, the wreckage points to the possibility that the mechanism was not "adequately lubricated." Part of the investigation also is addressing questions concerning the effectiveness of maintenance procedures and the oversight of maintenance by the Federal Aviation Administration (FAA).

The only explanation for the airplane's fatal dive, which is provided by the preliminary analysis of the data from the jet's flight recorder, is that the jackscrew assembly's end stop broke off in flight. The end stop, which is attached to the bottom of the jackscrew, is part of the flight control mechanism and is designed to assist with the proper operation of the plane's horizontal stabilizer. The stabilizer, which resembles a wing, is part of the tail structure and is used to control the plane's ascent and descent, or to maintain level flight. As the jackscrew turns through a gimbal nut, it tilts the front edge of the stabilizer at different angles, pushing the nose of the airplane up or down. An upward tilt of the stabilizer pushes the nose of the airplane downward, and vice versa. The end stop prevents the stabilizer from tilting up more than 2.2 degrees, which results in the maximum downward force on the aircraft's nose. Based on the results of the performance analyses and a simulation study to replicate Fight 261's final moments, the NTSB now believes that the stabilizer actually may have tilted up 22 degrees, or 10 times the safe maximum. The only way for that to happen, or even to tilt the stabilizer half a degree higher than normal, is to separate or break off the end stop from the jackscrew.

The salvage crew recovered the jackscrew from the ocean floor with the gimbal threads wrapped around it like a Slinky and the end stop missing. The investigators continue to look into the role of Alaska Airlines' maintenance program and procedures as a cause or a contributing factor to the crash. One possibility is that the jackscrew was not adequately lubricated, either because mechanics were lax or because the grease that Alaska Airlines used did not work, causing excess friction that could have in turn accelerated the wear of the gimbal nut and caused the threads to strip. Alaska Airlines' lubrication schedule for the jackscrew was less frequent than that of other carriers. At the time of the crash, Alaska Airlines lubricated the jackscrew every 2,500 flight hours or approximately every eight months. Other airlines lubricate the mechanism as frequently as every 500 flight hours, or five times as often.

The investigation is probing into the possibility that the plane's jackscrew assembly was so worn that it would have required replacement three years before the crash. Records indicate that an Alaska Airlines mechanic had ordered the part to be replaced in September 1997 but was later overruled by other mechanics.

It does not seem likely that the design rather than the maintenance of the stabilizer system contributed to the crash. In 1960, in order to acquire certification of the MD-80 stabilizer system, McDonnell Douglas designers provided the FAA with analyses showing that the probability of the end stop's failure was one in one billion.

Although it was a sad incident, it is interesting to note the following findings that were made after the crash of the Flight 261—findings that have a

(continued on next page)

direct bearing on the role of maintenance and maintenance procedures.

After the crash of Flight 261, the FAA ordered an immediate inspection of planes with the same stabilizer control system. As a result of this order, 23 planes were found to have potentially dangerous jackscrew abnormalities. The discovery that so many planes were flying with potentially unsafe parts points to the fact that the federal airline safety regulatory system does not work as well as the public has the right to expect. It also suggests that the preventive and predictive maintenance procedures and programs, at least for some airlines, have much room for improvement and currently leave a great deal to be desired. No regulatory system and no maintenance program ever should have allowed those planes to be flying with defective or substandard parts.

It is also disturbing to learn that Alaska Airlines mechanics had decided three years earlier to replace the part but reportedly were overruled the *next day* after further testing of the part and without waiting for results of standard follow-up tests. Subsequently, how much attention did Alaska Airlines and its maintenance program pay to that stabilizer after it was flagged, then unflagged, for replacement?

Prior to June 1998, three airlines had reported eight corroded stabilizers on the MD-80 series and 20 cracked ones were found by the end of the same year. In June 1998, the FAA issued a directive that airlines disassemble the tail section and replace the damaged parts within 18 months. The cost? A total of 117 hours and $7,000 per plane. In the case of Flight 261, this would have translated to less than $80 per victim—a very small price to pay to save a life, or in this instance, 88 lives.

Sources

Acohido, B. "Flight 261 Probe Shifts to Boeing." *Seattle Times*, 2000. http://archives.seattletimes. nwsource.com/cgi-bin/texis/web/vortex/.

Associated Press. "Broken Parts Blamed for Plane Crash." 20 November 2000. http:// www.dailynews.com/archives/2000/11/20/ new04asp.

"FAA Regulators Soft on Airlines" [editorial]. *Seattle Post-Intelligencer*, 2000. http://www. slackdavis.com/alaska_feb.html.

Malnic, E. "Tape Shows Pilots' Fight to Control Doomed Jet." *Los Angles Times*, 2000. http:// www.latimes.com/news/state/reports/ alaska/.

Press release, 27 October 2000. Washington, DC: National Transportation Safety Board [SB 00-25].

Taylor, C. "Many Airlines Lubricated Jackscrews More Often than Alaska." *Seattle Times*, 2000. http://www.slackdavis.com/ alaska_feb.html.

Figure 1–4 further reveals the breakdown and the causes of U.S. civil aviation accidents as provided by the NTSB. These incidents point to maintenance, or more specifically the lack of adequate and proper maintenance practices, as a cause or a major contributing fac-

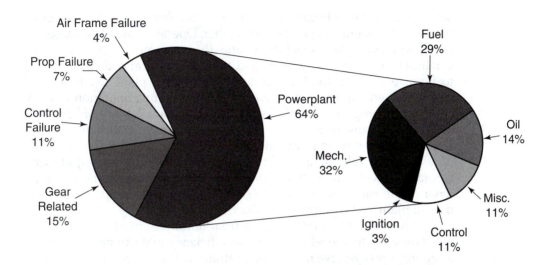

FIGURE 1–4   U.S. civil aviation accidents involving maintenance as a cause or factor

(Source: United States National Transportation Safety Board (NTSB), Washington, DC.)

tor. Once again, it is difficult to comprehend why these accidents continue to occur, given the available data, statistics, and knowledge of contributing factors, and the measures that can be taken to prevent such catastrophes.

1.8   CHANGING MAINTENANCE STRATEGIES

As discussed previously, optimizing the efficiency of assets has a significant impact on quality and productivity and affects the overall profitability of an organization. Often, this may determine whether or not the organization has the ability to compete or even survive. Although in most cases management may realize the benefits that can be gained from a well-executed asset management program, implementation of such programs usually requires a significant shift in the culture of the enterprise.

Maximizing the integrity and efficiency of a plant and its equipment requires a dramatic shift from the traditional maintenance philosophy and practice, often reactive in nature, to a proactive, well-planned process that is fully integrated across the plant. This change emphasizes the overall equipment effectiveness and integrity.

Operator-driven reliability is an important and ever-increasing aspect of this shift in maintenance strategy. Operator-driven reliability is a simple concept that brings the company's reliability maintenance practices and the operation personnel together. Operators are those closest to equipment and plant machinery, and they are most sensitive to the smallest changes in equipment conditions. It is logical and cost effective to involve these individuals as key resources in plant reliability and equipment effectiveness improvement efforts. Implementation of basic educational programs coupled with the current availability of handheld and other portable technologies provide operators with the ability to collect, store, monitor, and evaluate real-time equipment conditions and to play a major role in assuming ownership of the plant equipment. Operator-driven reliability closes the gap between operations and maintenance and can open the door to achieving significant improvements in overall plant and equipment productivity.

Figure 1–5 illustrates the path from minimum production and service efficiency to a level of maximum efficiency that can be achieved by adopting a progressive maintenance strategy. The figure shows the wide range of maintenance practices, from an ineffective reactive maintenance to an integrated world-class maintenance planning strategy.

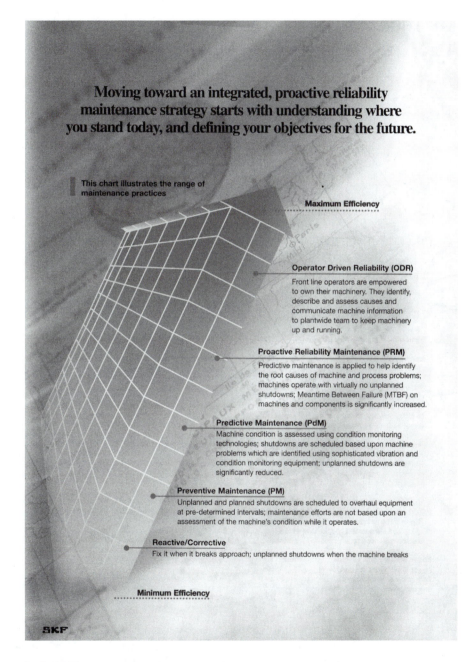

FIGURE 1–5   Moving upward in maintenance planning strategy
(Source: Courtesy of SKF Reliability Systems.)

QUESTIONS

1. How does the complexity of modern manufacturing systems affect maintenance planning and management?

2. Define:
   a. maintenance
   b. failure

3. What are the three types of maintenance activities and how do they differ from each other?

4. What does it mean to say that the maintenance department should be treated as a "profit center"?

5. How does the level of maintenance commitment affect preventive maintenance and repair costs?

6. State and briefly explain some of the primary objectives of the maintenance department.

7. State and briefly explain some of the secondary functions of the maintenance department.

8. In planning maintenance activities, what are some factors that the scheduler must consider?

9. What is meant by "maintenance backlog," and would you recommend keeping the backlog at or near zero level? Explain your answer.

10. Calculate the crew size for a maintenance department in order to maintain a backlog of four weeks if 1,500 hours of maintenance activities are scheduled per week. Assume a 35-hour workweek.

11. Calculate the backlog for a 15-person maintenance department. Assume a 40-hour workweek and 1,200 labor hours of maintenance activities.

12. Define CMMS and explain the importance and the roles of CMMS in modern maintenance management.

13. What is TPM?

14. Who is the primary focus of the TPM philosophy? Explain your answer.

15. TPM programs strive to reduce and eliminate six significant obstacles to achieving full equipment effectiveness. State and briefly explain the obstacles and how TPM addresses each of them.

16. What are the advantages and disadvantages of centralized versus decentralized maintenance departments?

17. In your opinion, could the crash of Flight 261 (from the case study) have been prevented? How?

18. Can you justify preventive maintenance in all situations and at any cost? Explain.

19. In your experience and from personal observations (your home, residence hall, work environment, automobile), what levels of preventive and predictive maintenance activities have you noticed?

20. What are some of the service industries that affect your daily life? How important are the various maintenance functions in the operation of these services?

2

STATISTICAL APPLICATIONS

Overview

Objectives

At the completion of the chapter, students should be able to

- Understand and define *reliability*.
- Use the equation provided to find the failure rate.
- Understand how to improve system reliability.
- Explain the various stages of the life-cycle failure rate.
- Develop an understanding of Weibull probability distribution and its significance to the study of reliability.
- Use graphical and mathematical methods to determine MTBF and make decisions regarding PM intervals and acceptable risk levels.
- Understand the concepts of queuing theory and applications.
- Describe various queuing models.
- Perform basic queuing model calculations.

Developing a basic knowledge of statistical tools and concepts of probability distributions is an important step in understanding certain fundamental theories for planning a sound preventive and predictive maintenance program. Essential reliability, queuing (waiting line), and normal and Weibull (exponential) probability distribution theories and applications are presented in this chapter.

2.1   RELIABILITY

Simply stated, reliability is the ability of a system to perform its intended function during its expected life period. In other words, a machine, component, or product, over its expected life period, should be able to perform its function at its expected level of capacity. A more technical definition of reliability states that reliability of a system or a product is the *probability* that the system will perform its specified function under the specified conditions throughout its specified life expectancy.

The key word in the definition is *probability*. Therefore, whether the system will perform its intended function is not a certainty but is a matter of chance and random occurrences. The function of a maintenance program then can be expected to include all activities that will keep the system in optimum working order during the life of the system. Whereas an appropriate preventive and predictive maintenance program can improve and enhance these chances by contributing to the reliability of a system, it is equally important to understand the statistical reliability of the design of the system in order to plan for an adequate maintenance program.

2.2   SYSTEM RELIABILITY

Most operation equipment or facilities are treated as a system rather than as individual components. A system is a logical collection and arrangement of a series of components working as a whole in order to perform a specific function. In order for the system to achieve its objectives, each component must perform its function in harmony with the entire system. It stands to reason that each component is in itself a mini-system, also comprised of a series of internal components. A simple example of a factory system includes a variety of equipment and machinery and so on, working as an integrated whole to achieve the production objectives. If one of these units fails, the overall system can fail.

The overall reliability of a system is a function of the number of components, the configuration or arrangement of these components in the system, and the reliability of each individual component. The system components can be arranged in series, parallel, or a combination of the two.

2.2.1   Series Systems

When components of a system are arranged in series, the system reliability is the product of the reliability of each individual component, assuming that the reliability of each component is independent of the reliability of the other components.

Series Reliability Equation

$$R_s = R_1 \times R_2 \times R_3 \times \ldots \times R_n$$

Figure 2–1 illustrates a system composed of four components in series. If the reliabilities of these components are given as 0.85, 0.90, 0.99, and 0.80 respectively, then the overall reliability of the system can be calculated as follows:

$$R_s = 0.85 \times 0.90 \times 0.99 \times 0.80$$

$$= 0.61$$

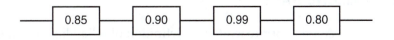

FIGURE 2–1   Components arranged in series

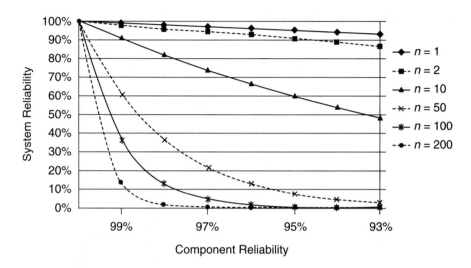

FIGURE 2–2   System reliability as a function of number of
components

The system reliability is approximately 61%. As stated earlier, reliability
represents a probability. A reliability of 0.61, or 61%, means that the system will perform its function nearly 61% of the time. Stated differently,
there is a 39% chance that the system will *not* operate, or will fail.

As the number of components in a series increases, the overall reliability of the system decreases rapidly. Assume three components, each
having 0.95 reliability, are connected in series. The system reliability is

$$R_s = (0.95)^3$$

$$= 0.86$$

Doubling the number of components, each with the same reliability, decreases the system reliability to 0.81. Figure 2–2 illustrates overall system
reliability as a function of the number of components and component reliability when components are placed in series.

2.3   FAILURE RATE

Another approach to the definition of reliability is the absence of failure.
Or, conversely, failure is the absence of reliability. Therefore, the reliability of any system is a function of its failure rate, λ. If the failure rate, or

failures per a given period of time, is stated as λ, then the system reliability can be calculated as follows:

$$R = 1 - \lambda$$

2.3.1   Determining the Failure Rate

Failure rate, λ, is the ratio of the number of failures to the total number of item-hours tested. Failure rate data can be obtained from product manufacturers or it can be determined from life tests and historical data.

$$\lambda = \frac{\text{Number of failures}}{\text{Number of item - hours tested}}$$

Assume that seven units of a particular product are tested for a period of 80 hours. If three of the units fail after 15, 35, and 60 hours respectively, calculate the product failure rate, λ.

The number of failed items is 3. In order to calculate the denominator, keep in mind that three units operated for 15, 35, and 60 hours respectively, before they failed. The remaining four units were still operational at the end of the test period.

$$\lambda = \frac{3}{15 + 35 + 60 + (4 \times 80)}$$

What is the reliability of this product?

$$R = 1 - \lambda$$
$$R = 1 - 0.007 = 0.993$$

Note that the reliability obtained here is the product reliability at the first hour of operation or at the time $(t) = 1$. As we will see later in this chapter, reliability will deteriorate as a function of time.

2.4   MEAN TIME BETWEEN FAILURES, AVAILABILITY, AND MEAN DOWNTIME

An important concept in reliability, useful for preventive and predictive maintenance planning, is mean time between failures (MTBF). Equally important are the associated issues of equipment availability and downtime. These concepts are explained in the following sections.

2.4.1   MTBF Defined

MTBF is the expected average time or the expected frequency with which we can expect the equipment to fail. Based on historical data and statistical probability, MTBF for a given component or system is the time interval during which we can expect the unit to perform its function after installation, proper maintenance, or overhaul. Although MTBF (mean time between failures) and MTTF (mean time to failure) are often used interchangeably, a clear difference exists between the two. If the failure of a unit is temporary and the unit can be repaired and brought back online by fixing it or replacing specific components, then MTBF is the applicable term. MTTF is therefore applicable to units or components that are disposed of at failure and are totally replaced. Hence, MTTF can be equated simply to the life expectancy of a unit.

Equipment availability is the proportion of time that we can expect the equipment to be up and operational. The average total amount of time that it takes to return failed equipment back online and ready for operation, from issuing a work order, to dispatching maintenance crew, to completion of the tasks, is referred to as the mean downtime (MDT).

2.4.2   Calculating MTBF

Mean time between failures is the reciprocal of failure rate:

$$\text{MTBF} = \frac{1}{\lambda}$$

Therefore, for the previous example,

$$\text{MTBF} = \frac{1}{0.007}$$

$$\text{MTBF} = 143 \text{ hours}$$

Therefore, within the laws of probabilities, we can *expect* this equipment to fail after approximately 143 hours of operation. Once again, this is *not* a certainty but a likelihood, a chance. This estimate gives maintenance planners something to work with, rather than attempting to determine the frequency of PM schedules without such data.

Given the MDT, derived from historical data and experience, and MTBF, equipment availability is calculated as follows:

$$\text{Availability} = \frac{\text{MTBF}}{\text{MTBF} + \text{MDT}}$$

2.5   IMPROVING SYSTEM RELIABILITY

Improving system design and adding redundancy or backup can increase system reliability. In a system with redundancy, when the first component fails, the backup component will keep the system operational and system failure is avoided. In order to add redundancy and increase system reliability, the components are placed in *parallel*.

2.5.1   Parallel Systems

For the system to fail in a parallel design, all components that are in parallel must fail. Figure 2–3 shows two components with reliabilities of 0.85 and 0.90 respectively, in a parallel design. Reliability of a parallel system is calculated as shown in the following equation.

Parallel System Equation

$$R_p = 1 - (1 - R_1)(1 - R_2)\ldots(1 - R_n)$$

$$= 1 - (1 - 0.85)(1 - 0.90)$$

$$= 0.985$$

Notice that the overall reliability of the system is *greater* than the reliability of each individual component.

2.5.2   Combinations of Series and Parallel Systems

Most systems are a combination of both series and parallel arrangements. In order to calculate the reliability of these complex systems,

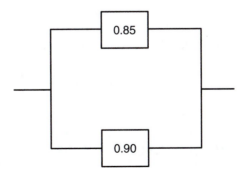

FIGURE 2–3   A system with two parallel components

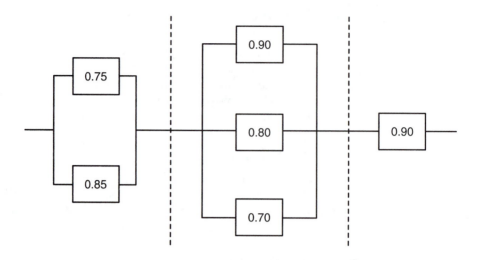

FIGURE 2–4   A system with some built-in redundancy

reduce the system to its basic structure. Figure 2–4 illustrates a combination of series and parallel arrangements. The solution to the example follows.

The reliability of the first section is

$$R_p = 1 - (1 - 0.75)(1 - 0.85)$$
$$= 0.9625$$

The reliability of the middle section is

$$R_p = 1 - (1 - 0.90)(1 - 0.80)(1 - 0.70)$$
$$= 0.994$$

The overall reliability of the system is

$$R_s = (0.9625)(0.994)(0.90)$$
$$= 0.86$$

2.6   EQUIPMENT LIFE CYCLE FAILURE RATE

For most products and components, failure occurs at different rates during the life of the product and follows different statistical and probability distributions. Figure 2–5 shows the failure rate of a product as a function

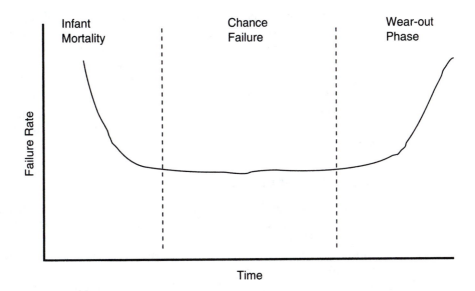

FIGURE 2–5 Product life cycle or bathtub curve

of product lifetime. The curve is known as the life cycle or failure rate curve. Because of its distinctive shape, it is also commonly referred to as the "bathtub" curve.

2.6.1 Infant Mortality

The first stage of the curve shows a high failure rate, which rapidly decreases as a function of time. The initial high failure rate, known as infant mortality or the debugging stage, is due in large part to improper design or manufacture, defective or inadequate components, or improper installation. For a diesel engine, for example, this phase is 1,000 hours of operation. Most manufacturing defects are discovered during the first 20 hours. The next 980 hours of operation usually reveal both manufacturing and design defects. Other factors, including improper calibration or use, also may result in product failure during this early stage of the product life. Borrowing the term from the field of electronics, this phase of the product life is also known as the "burn-in" phase. During this stage of the product life, the Weibull probability distribution (discussed later in this chapter) is applicable in most cases.

2.6.2 Chance Failure

The failure rate during the second stage of the life cycle follows a near-constant rate. This phase of the product or equipment life constitutes the useful or the productive stage of the product life. Failures during this phase are random or chance failures, which occur in a random manner. The useful life or the chance failure phase of the product is the segment of the product life cycle that is most important in preventive maintenance planning. During this stage of the equipment life, data to determine MTBF are collected and analyzed. For most products, the assumption that the failures occur at a constant rate is quite valid, and therefore, the exponential probability distribution is used during this phase of the product life.

2.6.3   Wear-out Phase

The final phase of the life cycle curve is the wear-out phase of the equipment life. The failure rate rapidly increases as a function of time during this stage. By this point most equipment is probably fully depreciated, and now becomes a candidate for retirement, for use as backup, or may be cannibalized for spare parts. As long as it can be economically justified, the equipment's useful life can be extended and a proper and a proactive preventive and predictive maintenance program can delay the wear-out phase. In most cases the normal probability distribution best describes this phase of the product life cycle.

It is imperative to realize that not only do various systems exhibit distinctly different failure rates, but each component within a given system also possesses a characteristic bathtub curve. Figures 2–6a, 2–6b, and 2–6c show bathtub curves that are characteristically typical of mechanical, electrical, and software systems respectively. Figure 2–7 displays a composite bathtub curve showing the failure rate of each component and the overall bathtub curve for a diesel engine.

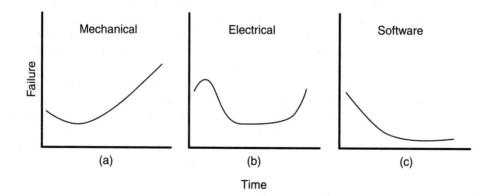

FIGURE 2–6   Characteristic failure rate curves

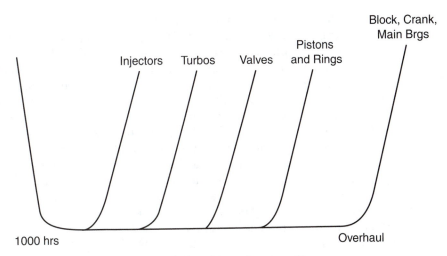

Injectors   Turbos   Valves   Pistons and Rings   Block, Crank, Main Brgs

1000 hrs   Overhaul

All components do not have the same life

FIGURE 2–7   A composite bathtub curve

2.7   EXPONENTIAL PROBABILITY DISTRIBUTION

As stated earlier, the useful stage of the equipment life is of most interest for a preventive maintenance program. Based on the life history data collected during this stage, the MTBF and other product behavior patterns can be determined and can be used to make important decisions regarding product maintenance, reliability, design modifications, warranty policies, and so on. For most components and products, the failure rate during this phase is constant and follows a random pattern. Exponential probability distribution can be used to determine product failure rate and predict its reliability.

2.7.1   Weibull Distribution

Exponential probability distribution is a special case of the more general probability distribution called the Weibull distribution. Swedish statistician and engineer Waloddi Weibull introduced the Weibull distribution to accommodate constant, increasing, and decreasing failure rates. It is important to realize that not all products exhibit a constant failure rate during this stage of their life cycle. As shown in Figure 2–8a, the product failure rate may increase, decrease, or remain constant as a function of time.

Therefore, the Weibull distribution, instead of the exponential probability distribution, would afford a more general treatment of the data during the useful life stage. Since the constant failure rate is the most common phenomenon, however, the special case of the Weibull distribution, the exponential probability, will be introduced here. As stated earlier, a careful examination of the data and the graphs is warranted in order to proceed with the correct analysis, however. Figures 2–8b and 2–8c illustrate this point. These probability plots are based on simulated data. Figure 2–8b displays a four-way probability plot based on simulated life test data with a constant failure rate; hence, the data best fit the exponential probability plot. Compare this figure with Figure 2–8c, which is based on simulated life test data with a nonconstant (either increasing or decreasing) failure rate. The data in this case best fit the Weibull plot, as shown in Figure 2–8c. These graphs, Figure 2–8b in particular, highlight the important fact that determination of the exact probability distribution at times may require additional knowledge of the process or even statistical and curve fitting procedures. These techniques, however, are beyond the scope of this discussion.

The main questions when planning a preventive maintenance program are when and at what intervals should a planned maintenance activity take place. Equipment must be first grouped into more or less homogeneous groups and then relevant data are collected for each group. Data should provide pertinent information in an unbiased manner about the type of maintenance work required, the frequency of breakdowns, and the cause of each failure.

The abbreviated breakdown history for a bank of process equipment is presented in Table 2–1. This table shows the length of operation before failure in hours for each machine in a group of similar and homogenous series of equipment. Since the data are collected during the useful life phase of the equipment, it is assumed to have a constant failure rate. The validity of this assumption will be tested during the data analysis steps, however.

2.8 GRAPHICAL ANALYSIS

The first step in data analysis is to create two additional columns, the cumulative number of failures and the cumulative percent failure, as shown in Table 2–2. The cumulative number of failures column is the result of the sequential summation of the number of failed units. In this example, one unit failed after 10 hours; after six more hours of operation, another unit failed, bringing the total of failed units to two after a total of 16 hours of operation; a total of three units failed after a total of 22 hours

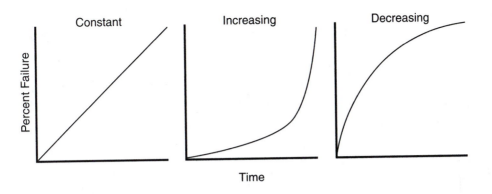

FIGURE 2–8a Constant, increasing, and decreasing failure rates

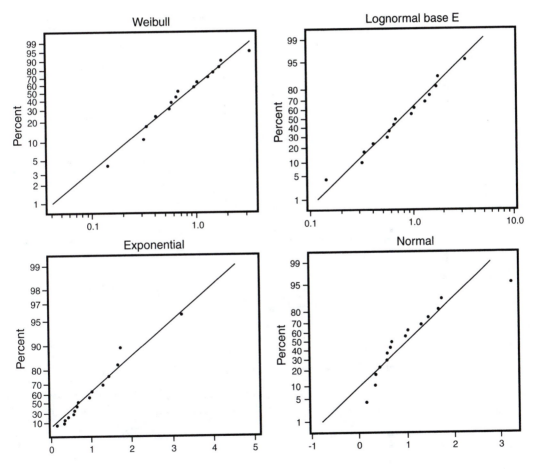

FIGURE 2–8b A four-way probability plot for life test data with constant failure rate (exponential fits best)

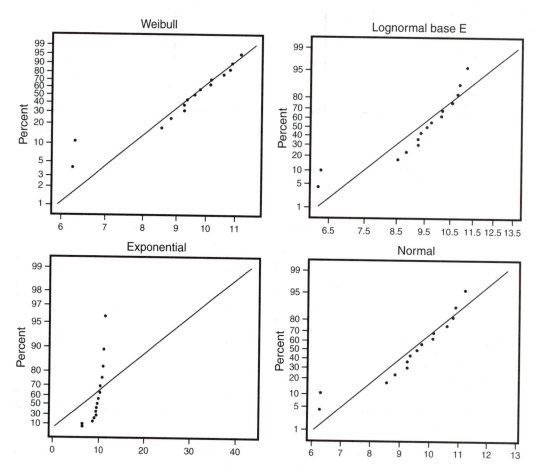

FIGURE 2–8c   A four-way probability plot for life test data with nonconstant failure rate (Weibull fits best)

TABLE 2–1

Equipment Failure History Data: Operation Hours Before Failure

Equipment ID#	Hours before failure	Equipment ID#	Hours before failure
473	10	479	56
477	16	457	75
453	22	472	98
459	32	456	140
478	43	458	180

TABLE 2–2

Equipment Failure Data with Cumulative Frequency and Percentages

Equipment ID#	Hours before failure	Cumulative number of failures	Cumulative % failure
473	10	1	10
477	16	2	20
453	22	3	30
459	32	4	40
478	43	5	50
479	56	6	60
457	75	7	70
472	98	8	80
456	140	9	90
458	180	10	100

of run time; and so on. The cumulative percent failure column keeps track of the total failed units as a percentage of the total units tested:

$$\text{Cumulative percent failure} = \left(\frac{\text{Cumulative number of failures}}{\text{Total number of test units}} \right) \times 100$$

After the failure of the first unit (one of the 10 units being tested):

$$\text{Cumulative} = \left(\frac{1}{10} \right) \times 100$$

$$= 10\%$$

Using exponential graph paper, shown in Figure 2–9, we can plot the time to failure (x-axis) versus the cumulative percent failure (y-axis). After plotting the points, construct a line of best fit through the points. Sometimes it may be necessary to ignore the first and the last points if they appear to be "outliers." The completed graph appears in Figure 2–10.

2.8.1 Obtaining Reliability Data from a Graph

Examination and analysis of the graph of the data reveals some important information regarding the product failure rate, MTBF, and product reliability. The use of exponential probability paper tests the validity of the constant failure rate assumption. The assumption is valid only if the resultant line is a *reasonably* straight line. If the line is not a straight line

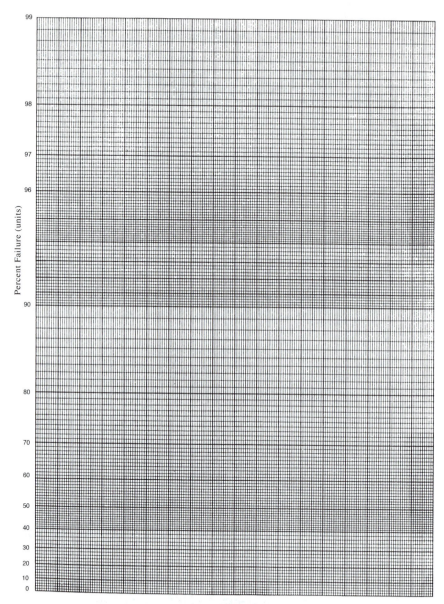

Age at Failure (hours)

FIGURE 2–9 Exponential graph paper
(Source: From Graph Paper from Your Computer or Copier *by J. S. Craver.)*
Copyright © 1996 by J. S. Craver. Reprinted by permission of Perseus Books, L.L.C.

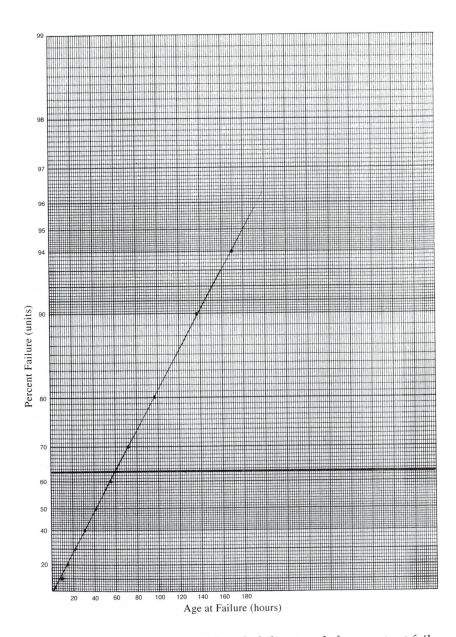

FIGURE 2–10 An exponential probability graph for constant failure rate on exponential probability paper

(Source: From Graph Paper from Your Computer or Copier *by J. S. Craver. Copyright © 1996 by J. S. Craver. Reprinted by permission of Perseus Books, L.L.C.*

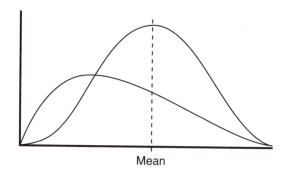

Mean

FIGURE 2–11   Location of mean as a measure of central tendency
                 for normal and Weibull distributions

and it has an upward or a downward tendency, as was illustrated in
Figure 2–8a, then the assumption of a constant failure rate is no longer a
valid assumption, and use of the Weibull distribution is warranted. Once
the assumption of a constant failure rate is satisfied, the plot may be used
for further analysis.

It is necessary to stress that, unlike the normal probability distribution,
exponential, and its more general probability distribution, Weibull, are *not*
bell-shaped, symmetrical distributions. These distributions are positively
skewed and the mean of these distributions is not the same as the median,
or the 50th percentile. In fact, as shown in Figure 2–11, the mean of a Weibull
distribution is located at the 63.2 percentile. This remains true whether we
are dealing with a constant, increasing, or decreasing failure rate.

We can conclude that the MTBF for an exponential distribution will
occur at the 63.2 percentile, not the expected 50th as with a normal probability distribution. If we draw a line perpendicular to the y-axis at the
63.2 percentile, the line will intercept the failure rate line at the MTBF
point. Therefore, 63.2% of the equipment will fail *before* the MTBF, which
is an extremely important consideration in setting the preventive maintenance schedule in relation to the MTBF time.

The 63.2 percentile failure rate line intercept shows the MTBF as
approximately 62 hours. Mathematical calculations of the MTBF confirm
our estimation:

$$\lambda = \frac{\text{Number of failures}}{\text{Number of item - hours tested}}$$

$$= \frac{10}{10 + 16 + 22 + 32 + 43 + 56 + 75 + 98 + 140 + 180}$$

$$= 0.015 \text{ failures per hour}$$

$$\text{MTBF} = \frac{1}{\lambda}$$

$$= \frac{1}{0.015}$$

$$\text{MTBF} = 66.67 \text{ hours}$$

2.8.2   Determining the Desired PM Interval from a Graph

The plot also can be used to determine desired intervals for a preventive maintenance schedule. Given the equipment failure data, several alternatives are available to the decision makers. The company may decide that they can afford *not* to have a preventive maintenance program and let the equipment fail before any actions are taken to fix the equipment. In this case, the time between failures will follow a pattern very similar to the *skewed* probability distribution in Figure 2–11. Approximately 63.2% of the equipment will break down after a few hours but before the MTBF, whereas some equipment (only 36.8%) may continue to operate well beyond the MTBF.

A second alternative would involve performing a preventive maintenance program after 65 hours (approximate average time between failures). In this case, the probability exists that 63.2% of the equipment will fail before any preventive action is taken.

A third course of action is to plan for preventive maintenance activity on a set interval that would offer the most acceptable failure risk. Let us assume that a 30% failure rate presents an acceptable risk factor, given other economic considerations. A line from the 30% mark intercepts the failure rate line at approximately 22 hours of operation. So if the company schedules a maintenance routine after 22 hours of operation, only 30% of all the equipment would fail or a specific piece of equipment would fail only 30% of the time before the scheduled maintenance routine. Since reliability was defined as the absence of failure, then 30% failure would translate to 70% reliability. The graph can be used to estimate any level of failure (or reliability) as a function of time intervals or vice versa—for a desired level of reliability, we can estimate the maintenance intervals.

2.9   MATHEMATICAL ANALYSIS

Although graphical analysis provides a simple means of determining product failure rates and can be used to estimate preventive maintenance intervals in order to attain desired system reliability, mathematical formulas can provide faster and more accurate answers. Once we have established that the product failure rate is constant, we can determine product reliability as a function of time and calculate more accurate maintenance intervals based on acceptable risks and desired reliability levels.

2.9.1   Failure Rate Equation

The failure probability density as a function of time can be written as

$$f(t) = e^{-\lambda t}$$

2.9.2   Reliability at Time (t) Equation

It follows that for a constant failure rate

$$R(t) = e^{-\lambda t}$$

where $R(t)$ is the expected reliability at time t for a constant failure rate λ.

In order to illustrate the use of the formula, we will calculate the product reliability at various time intervals using previous data. First, what is the system reliability at time $t = 1$?

$$R (1) = 2.7183^{-(0.015)(1)} = 0.985$$

Recall that this is the same value obtained previously from the expression $R = 1 - \lambda$, and we noted that this corresponded to reliability at the beginning of operation, or $t = 1$.

Now let us calculate system reliability at $t = $ MTBF $= 66.67$ hours.

$$R (66.67) = 2.7183^{-(0.015)(66.67)} = 0.3678$$

The preceding discussion indicated that the probability of failure for an exponential distribution at MTBF is approximately 63.21%, and since

$$R = 1 - \% \text{ failure}$$

$$R = 1 - 0.6321 = 0.3679$$

This is the value we obtained from the previous calculation.

The formula for calculating $R(t)$ can be used to determine preventive maintenance intervals in order to achieve a desired level of reliability based on acceptable risk (failure) levels. To illustrate this procedure, let us assume, as we did in the previous discussion, that a 30% failure can be tolerated as an acceptable risk factor, which translates to 70% system reliability. Assuming a constant failure rate of $\lambda = 0.015$, we can calculate (t) at which $R = 70\%$.

$$R(t) = e^{-\lambda t}$$

$$0.70 = e^{-0.015t}$$

$$\ln 0.70 = -0.015t \ln(e)$$

$$t = \frac{\ln 0.70}{-0.015}$$

$$t = 23.78 \text{ hours}$$

Once again, you may recall that based on our graphical estimation, a preventive maintenance routine after 22 hours of operation was required to reduce the failure risk to a 30% level or provide a 70% system reliability.

It is noteworthy to examine the outcome if we were to assume a normal distribution. Using the following formulas, the average is 67.2 hours and the standard deviation of the sample is equal to 56.7 hours.

$$\bar{x} = \frac{\Sigma x_i}{n} \qquad s = \sqrt{\frac{\Sigma(x - \bar{x})^2}{n-1}}$$

where $\bar{x}$ = the average time between failures

n = sample size

x_i = any value on the x - axis (hours)

s = sample standard deviation

It would follow that only 50% of the equipment would be expected to fail before the MTBF, not 63.2%. Furthermore, for 70% reliability (30% failure before maintenance), we would have calculated the maintenance interval, using the standard z-score formula, as:

$$z = \frac{x - \bar{x}}{s}$$

where z = standardized numerical value corresponding to the probability of an event

$$-0.52 = \frac{x - 67.2}{56.7}$$

$$x = 38 \text{ hours}$$

Maintenance would be scheduled after 38 hours of operation, not after 22 hours, and therefore the actual equipment reliability would have fallen short of the perceived or anticipated reliability.

2.9.3 Economics of Acceptable Risk Levels

Decision regarding acceptable risk levels are economical for the most part. Consider the following example. Assume the failure cost (emergency repair, lost production, and so on) for the equipment is $7,500 per occurrence. If the company institutes a preventive maintenance program to avert this particular type of failure, the cost would be $250 per scheduled routine. In order to decide whether or not to accept the PM program, the annual cost for each alternative (PM or no PM) is calculated. The following example serves as a means of illustrating these calculations.

Assume that the equipment is operated 260 workdays, 16 hours per day, for a total of 4,160 hours per year. With no maintenance option, the equipment is expected to fail, on the average, 4160/66 = 63.03 times (using MTBF = 66 hours) during the year, for an annual cost of 63.03 × $7,500 = $472,725. On the other hand, a PM program that would allow for 20% failure would require 4160/16 = 260 scheduled PM per year at an annual cost of 260 × $250 = $65,000. (Note: The 20% failure rate corresponds to 16 hours age at failure on the graph.) There remains a 20% probability that the equipment still may fail before the scheduled maintenance activity. Therefore, 260 × 0.20 = 52 annual failures are expected before PM at the cost of 52 × $7,500 = $390,000 per year, and the total cost for the PM option is $65,000 + $390,000 = $455,000 per year. In this scenario, the PM option offers a saving of $472,725 − $455,000 = $17,725 per year per machine. The savings then can be multiplied by the number of machines. It is important to keep in mind that these calculations are based on statistical probabilities and the law of averages.

The following case study illustrates the practical application of life cycle and reliability studies conducted by an international leader in diesel manufacturing. The data obtained as the result of such extensive studies are essential in determining and recommending the required maintenance types and intervals, product and various component life expectancies, and important warranty information. Careful attention to this study will bring in to focus the relevance and importance of these statistical techniques in manufacturing and industry.

A RELIABILITY STUDY FOR A DIESEL ENGINE

Manufacturers conduct life tests on their products and their components for a variety of reasons. These tests can determine the useful life of components and products, the mean time between failures, and product reliability. The data are used for warranty purposes, to establish preventive and predictive maintenance schedules and routines, to estimate product life cycle costs, for product liability protection, for quality and reliability improvements, and so on. Data obtained from such tests also are useful in design and process planning, evaluation, and improvements, and often are mandated by contracts, especially military contracts.

As discussed in the text, life testing involves continuous operation of several units of the products or components until the units fail. Based on the total of units tested, the total operating time for each unit, and the number of units still operational after a certain amount of time has elapsed, MTTF and other important product and component characteristics can be determined.

A leading international manufacturer of diesel engines routinely conducts life tests on various components and products. In the case presented here, life tests were performed on nine product models, with sample sizes varying from 17 to 755 units. More than 10 different components, including valves, turbos,

injectors, bearings, camshafts, cylinders, and so on, were tested.

The plot of the life test data is presented in Figure 2–12a. Some important information regarding the product failure rate, MTBF, and component reliability are obtained from the data. The use of exponential probability paper and the reasonably straight line of the plotted data confirm the validity of the constant failure rate assumption. Note that it was necessary to ignore the first and the last data points, as was mentioned in the earlier discussion about graphing the data.

It has been said that anything that cannot be measured cannot be improved. Product reliability can be improved only through proper data collection and analysis. Figures 2–12b and 2–12c show the plots of the life test data after further reliability studies and improvements. It is interesting to note that as the result of these studies, the failure rate of the manufactured products dropped from the initial 24.9% to 7.7%, which is a significant improvement.

Each component also exhibited a bathtub life cycle curve. (See Figure 2–7). The overall product life cycle curve is exhibited in Figure 2–13. As shown in the figure, the infant mortality or debugging stage lasts for approximately 1,000 hours of continuous operation, according to the test results. The straight line portion of the graph, which represents a constant failure rate during this period and lasts approximately 1,000,000 hours before the need for a major overhaul arises,

(continued on next page)

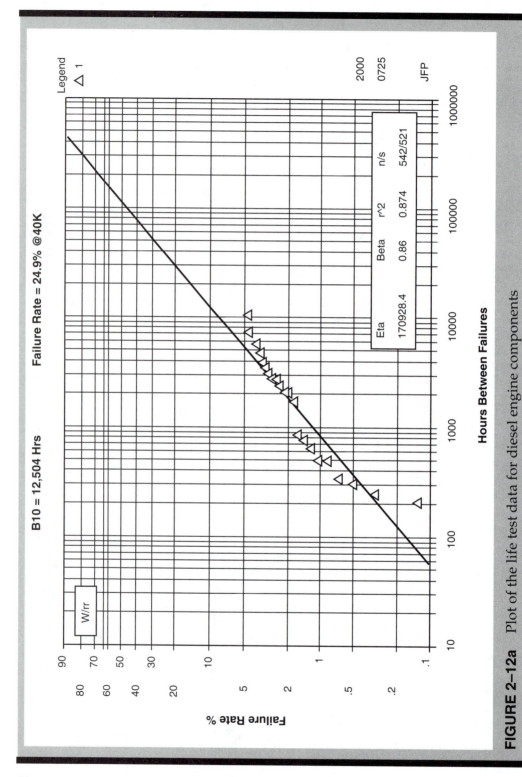

FIGURE 2–12a Plot of the life test data for diesel engine components

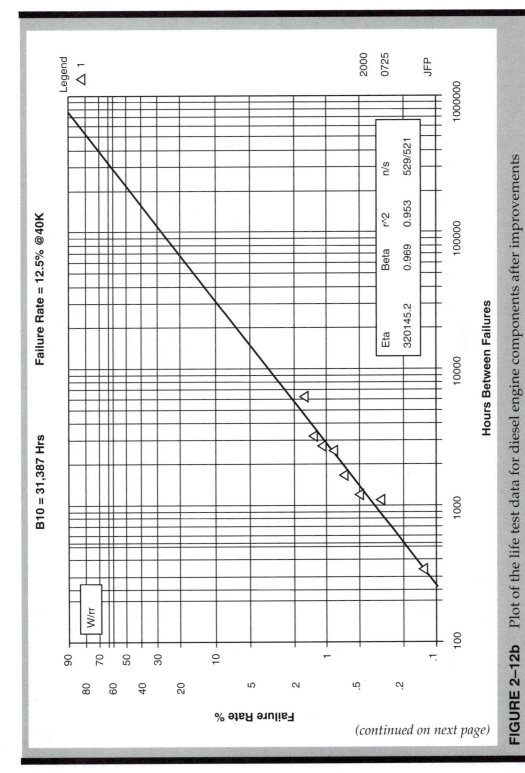

(continued on next page)

FIGURE 2–12b Plot of the life test data for diesel engine components after improvements

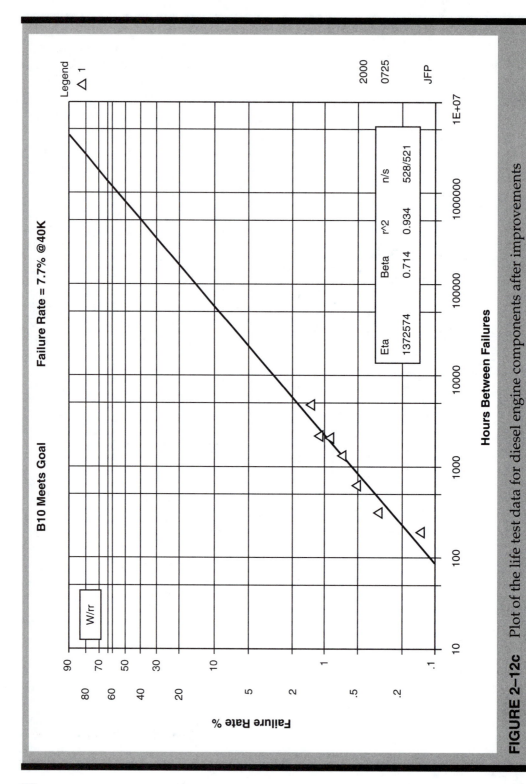

FIGURE 2–12c   Plot of the life test data for diesel engine components after improvements

depicts the useful stage of the product or component life.

The three additional graphs presented in Figures 2–14, 2–15, and 2–16 also are based on the results of the life test data. A careful and close study of each graph reveals different, yet interesting, information. Figures 2–14 and 2–15, which display warranty costs and the number of repairs respectively, in effect show the product or component failures during the first 20,000 hours of operation. The warranty costs and the number of repairs are quite high during the early stages of product life, which closely correspond to the debugging stage of the life cycle curve. However, as the normal or the useful life cycle begins, both warranty costs and occurrences of failures decrease and then maintain a constant rate.

The graph in Figure 2–16 displays the dollar cost per repair during the first 20,000 hours of operation. The graph shows an upward trend, indicating an increasing cost per repair after the debugging stage (the first 1,000 hours). Whereas the number of failures or repairs is relatively high during the debugging state, these initial failures are minor and most likely will require simple adjustments or alignments with low expenditures of time and other resources. Subsequent repairs, although fewer in number, are comparatively more major and often may require replacing parts or components; therefore they tend to be more costly per occurrence. This is another case that supports preventive maintenance: often minor and less expensive steps can help avert major and expensive problems from occurring in the future.

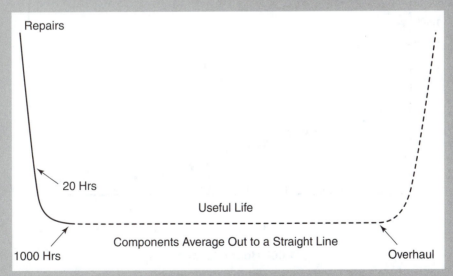

FIGURE 2–13   Bathtub curve for the overall product life cycle

(continued on next page)

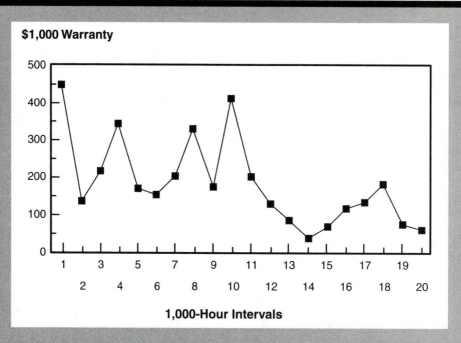

FIGURE 2–14 Plot of the warranty costs as a function of component life cycle

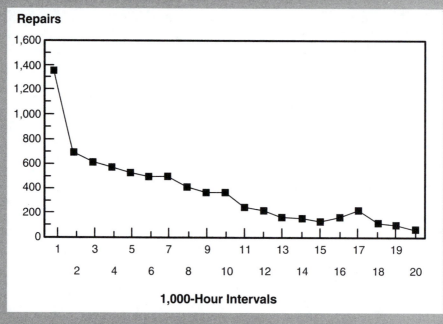

FIGURE 2–15 Number of repairs of diesel engine components as a function of their life cycle

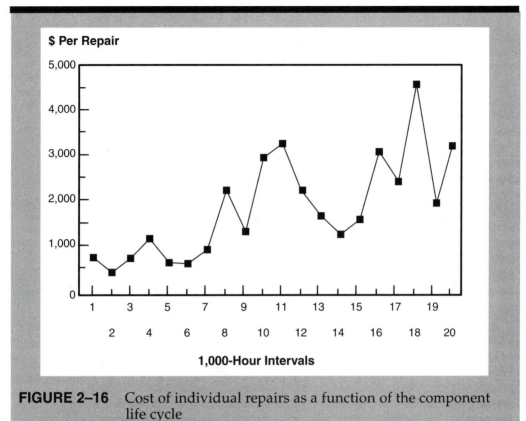

FIGURE 2–16   Cost of individual repairs as a function of the component life cycle

2.10 QUEUING THEORY AND APPLICATIONS

Maintaining and managing a proper backlog is an important factor in administration of a successful and a productive maintenance program. A large backlog or one that is continually increasing results in a slow response time and a deteriorating service level. The consequences of a small or nonexistent backlog include increased idleness of the maintenance personnel and diminished productivity.

Queuing or waiting line disciplines are applicable to all service activities. The waiting lines at supermarkets, fast food restaurants, banks, and the checkout counters at department stores are prime examples of queue or waiting line problems. Whereas long lines are indicative of a large number of customers, if the length of the line and the duration of the wait are not controlled, they can potentially result in the loss of business as customers seek better service elsewhere.

The obvious solution to a long waiting line at a bank, for example, may be to add additional tellers. Such an increase in the service capacity of the organization, however, comes with added costs for additional equipment, space requirements, and of course, the cost of added personnel. Therefore, the challenge facing the manager is balancing the needs of the customers with the extra allocation of resources.

Similarly, the maintenance department constitutes a service department and the equipment the department services forms a queue or waiting line for scheduled preventive maintenance, planned overhaul and modifications, or emergency repairs. Once again, it is the responsibility of the manager to strike a balance between the costs of the service level with the costs of the waiting time to arrive at the optimal service level.

2.10.1 Queuing Models

Quantitative queuing models have been developed to understand the behavior of waiting lines and enable the operation managers to make better decisions regarding the operation of these lines. Queuing models may be used in order to understand:

- **Input into the system**. These are the units, customers, equipment, and so on arriving at the queue and waiting for service. The population arriving at the queue may be of an infinite size, or may be finite and have a limited size. The cars arriving at one of the tollbooths in Chicago during the rush hour would be an example of an unlimited population arriving at a queue. In a moderate-size manufacturing facility, with a limited amount of equipment, the equipment that will fail and form a waiting line for service would be considered a limited or a finite population.

The arrival pattern of the units is another characteristic of the system inputs. In most queuing disciplines, arrivals follow the Poisson probability distribution and the number of arrivals per unit of time can be estimated as a number of units per hour or day. For planning a maintenance strategy, the rate of equipment failure constitutes the arrival pattern. Therefore, the rate at which the equipment fails, or the failure rate, is the rate at which the equipment arrives at the waiting line for repair.

The third characteristic of the arrivals is whether they remain in the line for service or get impatient and leave or jump to a different waiting line. Fortunately, in the maintenance queue, the arriving units do not have the luxury of becoming impatient or jumping to a different line, and therefore they are unable to complicate the model unreasonably.

• **Waiting line characteristics**. There are two major characteristics of waiting lines: the length of the line and the line discipline. A line is allowed to grow without bounds. Such a waiting line is said to be unlimited. The earlier example of the waiting line at a busy tollbooth is an unlimited queue. In the maintenance department, however, the waiting line or queue by necessity would be a limited or a restricted line. If such a line were to grow without bounds, it would mean that the maintenance department was getting farther and farther behind and all equipment would eventually be out of service and waiting for repair.

The queue discipline refers to the order in which the units in the line are serviced. Most lines follow the first-in, first-out (FIFO) procedure; that is, the customers or the units are serviced in the order of arrival. In some situations, however, it may become necessary for some priority arrangements to override the FIFO discipline.

• **Service characteristics**. The pattern of the service time and the configuration of the service facility are the major characteristics of the service system. For most activities, service time is random and is best described by a negative exponential probability distribution. Once the average service time has been established based on historical data or various work measurement methods, we can determine the probability that a unit in the system will require beyond a certain amount of time:

$$P(\text{Service time} > x) = e^{-ux}$$

where

$$x \geq 0, \text{ and}$$

$$\mu = 1 \,/\, \text{Average service time}$$

Let us assume that the average service time is equal to 30 minutes. What is the probability that an arriving unit will require more than one hour of service?

$$P\,(\text{Service time} > 60 \text{ minutes}) = 2.718^{-(1/30)(60)}$$

$$= 0.135 \text{ or } 13.5\%$$

There is 13.5% likelihood that an arriving unit will require more than an hour of service. Stated differently, the probability is $1 - 0.135 = 0.865$ or 86.5% that arriving units will require less than one hour of service. Therefore

$$P\,(\text{service time} \leq x) = 1 - e^{-ux}$$

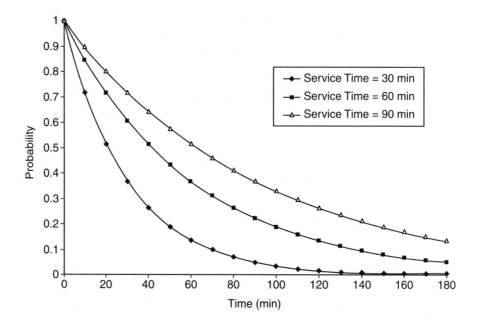

FIGURE 2–17   Negative exponential probability distribution

Figure 2–17 shows the probability of service for average service times of 30 and 60 minutes. Notice that for a given average service time, the probability that an arriving unit will require a very long service time is low. Given the average service time of 30 minutes, the probability that a unit will require more than 180 minutes of service is nearly zero (0.2%).

In most machine-operated service environments, such as an automatic car wash, service time is fixed and follows a constant pattern.

• **Service configuration**. The number of phases or interaction points between the arriving units and the service system and the number of servers determines the type of service configuration:

Single-channel, single-phase. This is the simplest queue configuration. In this setup, the customers or arriving units wait in a line or queue with only one server or service station. The units receive service from the server and then exit the system. The drive-in window at a bank (with only one drive-in facility) or a grocery checkout line with only one checkout clerk on duty are examples of this queue configuration, shown in Figure 2–18a.

Single-channel, multiple-phase. In this case, the customers interact with more than one server, as shown in Figure 2–18b. The bureau of motor vehicles, where various agents perform different tasks, or a fast food restaurant, where one clerk takes your order and receives payment and

Single-Channel, Single-Phase

(a)

Single-Channel, Multiple-Phase

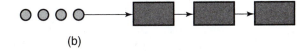

(b)

Multiple-Channel, Single-Phase

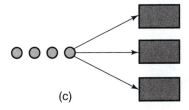

(c)

Multiple-Channel, Multiple-Phase

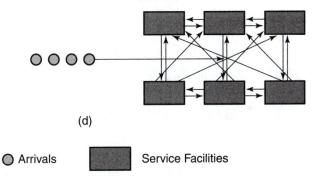

(d)

◯ Arrivals   ▮ Service Facilities

FIGURE 2–18   Various queue configurations

another serves your food, are common examples of single-channel, multiple-phase waiting line disciplines.

Multiple-channel, single-phase. A bank with several tellers on duty is an example of this system. A customer interacts only with the first available teller to receive complete service and then exits the system as shown in Figure 2–18c.

Multiple-channel, multiple-phase. Several steps or service phases are required for complete service, but multiple stations for each step of the service are available. Although the service steps are usually performed in a particular sequence, all stations are available to all customers and to further complicate the system, each phase may have its own waiting line. Figure 2–18d shows a multiple-channel, multiple-phase system in which arriving units may move or "jump" in a random manner to various service facilities in order to complete different stages of service or transactions.

The maintenance department of an organization constitutes a waiting line. Machines awaiting repairs or equipment in need of preventive maintenance form a waiting line or a queue. Maintenance managers, with the help of quantitative models, can understand and evaluate the performance of the maintenance department. Queuing models can help us understand the following operating characteristics:

- The average number of the units or customers waiting in the queue for service
- The average number of units in the system (waiting and being served)
- The average time each unit spends waiting in the queue
- The average time each unit spends in the system (wait and service time)
- The probability that a service facility is idle
- The probability that an arriving unit will have to wait
- The probability of a specific number of units in the system

2.10.2   Basic Queuing Model Calculations

In order to illustrate these calculations, we will use a numerical example. Since maintenance departments and their operations usually resemble a single-channel, single-phase system, we will limit our mathematical model to this particular configuration.

Let us assume that the equipment, either because of breakdown or for expected planned maintenance, enters the waiting line at the rate of three per day. This is called the arrival rate, λ (recall the failure rate). The maintenance department services the equipment on a FIFO basis. The average repair time is two hours. Given an eight-hour day, the service rate, μ, is $8/2 = 4$ per day. Therefore

$$\text{Service rate, } \mu = 4 \text{ per day}$$

$$\text{Arrival rate, } \lambda = 3 \text{ per day}$$

1. What is the average number of units waiting to be serviced or repaired?

$$L_q = \frac{\lambda^2}{\mu(\mu - \lambda)}$$

$$= \frac{3^2}{4(4 - 3)}$$

$$= 2.25 \text{ units}$$

We can expect that at any given time, on the average, 2.25 pieces of equipment are waiting to be repaired.

2. On the average, what is the total number of units that are either being repaired or are waiting to be repaired?

$$L_s = \frac{\lambda}{\mu - \lambda}$$

$$= \frac{3}{4 - 3}$$

$$= 3 \text{ units}$$

At any given time, an average of three units are out of service. These units are either being repaired or awaiting maintenance work.

3. What is the average wait time in the line?

$$W_q = \frac{\lambda^2}{\mu(\mu - \lambda)}$$

$$= \frac{3}{4(4 - 3)}$$

$$= 0.75 \text{ or } \frac{3}{4} \text{ of a day (6 hours)}$$

On the average an arriving unit has to wait nearly six hours *before* any repair work can start. This is the average wait time in the queue.

4. What is the average total time in the repair shop?

$$W_s = \frac{1}{\mu - \lambda}$$

$$= \frac{1}{4 - 3}$$

$$= 1 \text{ day (8 hours)}$$

Equipment requiring repair or service, on the average, is out of service for an entire day, or eight hours. Part of this time (six hours) is spent waiting in the line, as shown in the previous calculation, and the actual repair time takes two hours on the average.

5. What is the probability that the maintenance department is idle?

$$p_0 = 1 - \frac{\lambda}{\mu}$$

$$= 1 - \frac{3}{4}$$

$$= 0.25 \text{ or } 25\%$$

The probability that no equipment is waiting for repair or routine maintenance, that is, the likelihood that the service department is idle, is 25%.

6. What is the probability that an arriving unit will have to wait in line for service?

From the previous calculations, it follows that there is a 75% probability that the maintenance department is busy. This is called the utilization ratio. In other words

$$\text{Utilization ratio, } p = \frac{\lambda}{\mu}$$

$$= \frac{3}{4} = 0.75 \text{ or } 75\%$$

There is a 75% likelihood that an arriving unit will have to wait for service.

7. What are the probabilities that a certain number of units are in the system?

This examines the chances that a specific amount of equipment is out of service, either being repaired or awaiting repair:

$$p_n = \left(\frac{\lambda}{\mu}\right)^n p_0$$

What is the probability that two units are in the system?

$$p_2 = \left(\frac{3}{4}\right)^2 (0.25)$$

$$= 0.14 \text{ or } 14\%$$

QUESTIONS

1. Calculate the reliability for the following systems:

a.

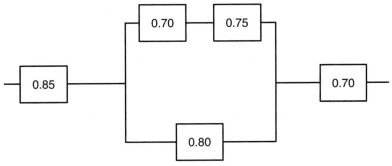

b.

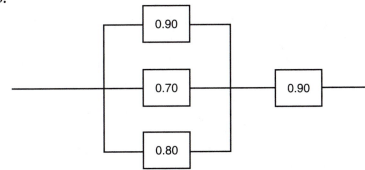

c.

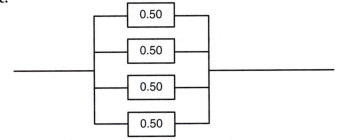

d.

2. Six components have been tested for a period of 90 hours. Four units failed after 20, 24, 45, and 65 hours respectively. If the average downtime is 30 hours, calculate
   a. the failure rate
   b. MTBF
   c. availability for this item
   d. reliability at the first hour of operation
   e. reliability at MTBF

3. Eight components have been tested for a period of 120 hours. Five units failed after 45, 55, 80, 105, and 110 hours respectively. If the average downtime is 60 hours, calculate
   a. the failure rate
   b. MTBF
   c. availability for this item
   d. reliability at the first hour of operation
   e. reliability at MTBF

4. Assuming an exponential failure rate and a MTBF of 250 hours, what is the item's
   a. failure rate
   b. reliability at 150, 250, and 350 hours of operation

5. Assume that the equipment is operated 300 workdays for 8 hours per day and with a known MTBF of 200 hours. Failure costs are estimated at $5,000 per occurrence. A PM maintenance option would cost $450 per scheduled routine. Could you justify the PM program if performing the PM routine every 120 hours reduces the probability of failure to 35%?

6. Assuming a constant failure rate of 0.025
   a. calculate $R(t)$, where $t = $ MTBF
   b. calculate the PM interval if only a 25% failure risk can be tolerated

7. For a constant failure rate of 0.002, calculate the PM interval for a 20% failure rate.

3
PREVENTIVE MAINTENANCE

Overview

Objectives

At the completion of the chapter, students should be able to

- Define *preventive maintenance*.
- Appreciate the significance of PM in productivity and profitability of an organization.
- Clearly understand the significance of maintaining equipment history.
- Understand how keeping an accurate equipment history and establishing a system of criticality play an important role in PM.
- Plan the various steps of preventive maintenance.
- Know where preventive maintenance starts.
- Calculate expected number of breakdowns, total failures, and total cost.

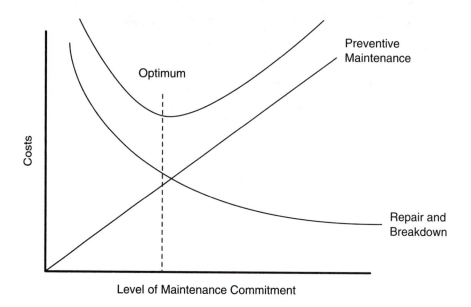

FIGURE 3–1   Comparison of various costs at different levels of
maintenance

Although some degree of preventive maintenance (PM) is absolutely necessary in developing and maintaining a reliable production system, economic considerations are essential criteria when instituting a feasible preventive maintenance program. As explained in the previous chapter and shown in Figure 3–1, with an increase in the level of preventive maintenance, the overall equipment effectiveness increases and breakdowns decrease. However, as the level and the degree of preventive maintenance increase, some of the costs associated with more PM activities will increase, and thus the overall costs may also increase. But, remember that some of these costs may be offset by decreased downtime, increased utilization of various resources, and an associated increase in productivity. Also, as illustrated in Figure 3–2, the costs associated with increased levels of preventive maintenance do not always increase linearly with the levels of preventive maintenance and may actually decrease. As organizations become more skilled in applying preventive and predictive maintenance techniques, as various personnel assume "ownership" of the processes and equipment, and as the initial costs of implementing preventive maintenance technologies are prorated to a larger scope of operations, the cost per unit of preventive maintenance activity likely will drop. This decrease, along with other benefits of preventive maintenance stated earlier, will significantly lower the overall

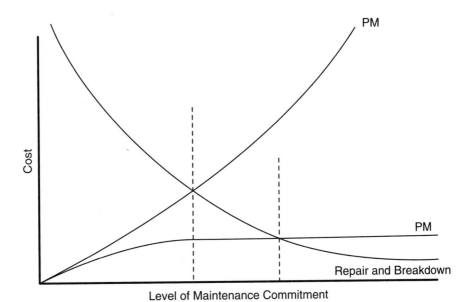

FIGURE 3–2   Maintenance costs comparisons with nonlinear PM

cost. This point is well illustrated in the lower PM plot line in Figure 3–2. Therefore, decisions about optimum PM levels should remain dynamic and will change as a function of the production environment, equipment conditions, and the criticality of the equipment.

A review of Table 3–1 brings into focus the prevailing levels of preventive maintenance activities practiced in American industries today,

TABLE 3–1

Current Preventive Maintenance Status in North America

Type of industry	Actual	World class
Assembly	29%	53%
Distribution	56%	54%
Manufacturing (Large)	29%	51%
Manufacturing (Small)	34%	52%
Process	34%	42%
Consultants' Opinion	25%	44%
Weighted Average	33%	47%

Source: Courtesy of Edward H. Hartmann, International TPM Institute, Inc.

compared with levels that would be practiced by a world-class organization. Table 3–1 shows that an ideal overall PM program consists of nearly 50% of an organization's total maintenance activities. So it is important to understand those activities that constitute preventive, as opposed to corrective or reactive, maintenance and determine how these programs can be successfully implemented in an organization.

3.1 PREVENTIVE MAINTENANCE

Preventive maintenance (PM) can be defined as a series of predefined and scheduled maintenance activities that are designed to reduce equipment breakdowns, increase equipment reliability, and improve productivity. The benefits of PM include increased equipment utilization and life, reduced work stoppage and machine slowdowns, closer adherence to production and delivery schedules, and increased employee morale.

Preventive maintenance activities can be grouped into two major categories, routine and major preventive maintenance.

3.1.1 Routine Preventive Maintenance

Routine PM is a series of clearly defined and routine activities that are normally carried out by the equipment operator. These activities allow for operators to participate in basic activities and should normally take only a few minutes to perform. Operators' involvement increases their sense of empowerment and equipment ownership and relieves maintenance personnel from performing routine tasks.

Routine PM must be performed systematically and according to a preset schedule. It may include such functions as cleaning, inspection, lubrication, and other minor repairs and adjustments.

Cleaning This includes removing dust, grease, and debris from equipment. Cleaning is the first step in the inspection and detection of any problem that might have occurred already or may develop soon. In addition to the esthetic value of clean equipment and the positive attitude that a neat environment can produce, dirt can cause heat, friction, product contamination, and electrical faults. During the cleaning process, operators increase their awareness of how the equipment operates, and their respect for the equipment increases. One of the functions of cleaning is making the equipment inspection easier. Therefore, isolating and treating the sources of dirt should enhance equipment condition.

Inspection All human senses, sight, hearing, smell, and touch, should be employed during the inspection. The operator should look for loose or broken parts, check for low levels of coolant, and notice any worn or frayed electrical wires. Loose or broken fasteners are a major source of vibration and breakdowns. Loose electrical connections create heat and thermal expansion and can cause fires and electrical shocks. Loose spaces collect dirt and cause oxidation. Routine inspections can detect disabled safety features, inoperative or malfunctioning gauges, lights, or other instruments, and leaks of hydraulics and other substances. Other problems to look for include visibly vibrating equipment or anything that seems to be abnormal. Listen for excessive and unusual sounds, rattles, and grinding noises, or hissing sounds that may indicate pneumatic and gas leaks. Listen for friction noises that may be due to low levels of lubrication or other problems. Finally, feel for vibration, excessive heat, and/or greasy surfaces that may indicate leakage and check for unusual odors, burning smells, and gas leaks.

Lubrication   The goal is to avoid having to follow the advice in the old adage about the squeaky wheel that gets the grease! The wheels should be greased *before* they begin to squeak. Historical evidence shows the existence of wheels dating to 3500 B.C. and point to our ancestors' concern with reducing friction. Any parts or surfaces in motion against each other require lubrication to reduce friction and wear. Understanding the importance of lubrication, increasing the ease of access to lubrication points, and reducing the types of lubricants by standardization will facilitate and encourage the practice. By combining the cleaning and lubrication functions, clogged or broken fittings can be detected, cleaned, and corrected.

Other routine functions   Other routine operator-based preventive maintenance activities may include testing the equipment, making minor adjustments, tightening loose bolts, and other simple, minor repairs. Do not underestimate operators' knowledge and understanding of the equipment. Operators' experience with and knowledge of the equipment can be an invaluable asset in keeping the equipment in perfect running condition as well as in detecting the first warning signal before a minor problem can turn into a major catastrophe.

3.1.2   Major Preventive Maintenance

Major preventive maintenance activities are performed by the maintenance personnel. These activities, compared with the routine PM activities, require longer periods of time, demand higher levels of skills, and usually necessitate scheduled equipment shutdown. These preventive measures aim to solve and correct the impending problems in an orderly and methodical fashion before the breakdown occurs. And since the shutdowns are planned, logically they will take place at a convenient time when all material, equipment, and skilled maintenance crew are available.

Major preventive maintenance activities may include partial dismantling of the equipment, major overhauls, or replacing worn parts and components. Major PM activities also include equipment upgrade and modification and often may require the involvement of the equipment manufacturer or vendor. Equipment installation, recalibration, and test runs are part of these activities.

The degree and extent of the preventive maintenance program required for a system can be based on the equipment history and a system of criticality.

3.2   EQUIPMENT HISTORY

No meaningful and effective preventive maintenance program can be established without a *documented* and clear understanding of the equipment history. We emphasize "documented," because relying on maintenance staff recollections to remember events associated with the equipment is as reliable as no history at all.

3.2.1   Establishing and Updating History

Establishing and updating an equipment history from cradle to grave, or more accurately from purchase and installation to salvage, is an essential part of the preventive maintenance and overall maintenance management program. It is important to establish a comprehensive equipment history for each piece of equipment showing purchase date, equipment identification number, vendor, and installation location. Original cost and all improvement costs must be kept as part of the equipment record. A record of all maintenance and repair activities, overhauls, rebuilds, and modifications must show dates and the extent of these activities. Specific and repetitive failures, possible causes, remedies, and corrective actions that were taken should be noted. These data are used to establish and further modify the preventive maintenance schedule.

3.2.2   Determining Reliability, MTBF, and Availability

Equipment history aids in determining equipment reliability, MTBF (mean time between failures), equipment availability, and mean downtime, as previously discussed. The data also can be used to evaluate existing maintenance programs and provide insight about how to modify current practices. Historical data are also essential for the analysis of failure trends to further refine and manage maintenance activities and to direct corrective actions. Analysis of the historical data can identify specific repetitive failures and their root causes. Steps can be taken to reduce premature equipment failures by design modifications or altering operational procedures.

The major source for building equipment history is the data generated from completed work orders. Reliable and complete historical data pave the way for predicting future behavior of the system and anticipating the problems that may occur. Equipment history data are used to determine equipment improvement needs and to compare maintenance costs with replacement costs. In general, work order and repair history can reveal a pattern of chronic problems and can signal the need for corrective action before serious damage is done to a machine.

Of course, it is prudent to realize that maintaining history for every piece of equipment is not warranted. Establishing equipment history is advisable only for situations in which the equipment is readily identifiable, preventive and predictive maintenance are desired, and the cost of data collection is justifiable and of value to engineering, accounting, maintenance, and other departments in the organization.

Figure 3–3 shows a basic equipment history data sheet, listing each maintenance or repair activity along with the associated and cumulative cost. A computerized data collection and retrieval system is helpful to maintain complete and accurate data.

3.3   ESTABLISHING A SYSTEM OF CRITICALITY

Of the hundreds and thousands of parts, components, and equipment in a facility, not all need a program of preventive maintenance. Even those that will benefit from preventive maintenance do not all require the same amount of PM. To apply the standards of a comprehensive PM program to every part and every piece of equipment would neither be cost effective nor even possible.

For a responsible and effective preventive and predictive maintenance program, it is important to establish a system that classifies the equipment based on the degree to which the equipment and therefore its failure is considered to be critical. This system of criticality can be based on several factors:

- **Equipment cost.** A piece of equipment priced at $250,000 certainly requires a greater level of maintenance than a $1,000 piece of equipment.

- **Function and redundancy.** A unique piece of equipment that can halt production when it malfunctions definitely requires closer

Equipment No. 021760

Installation Date August 27, 2003

Location Milling Center 7

Equipment Description Vertical Mill

Vendor ACME Corporation

Cost $65,000

Date	Work Order Number	Action Taken	Craft Code	Labor Hours	Labor Cost	Material Cost	Total Cost	Cumulative Cost
01/05/04	1722	Vibration Monitoring	02	1.0	25.00	—	25.00	25.00
01/17/04	1806	Change Filter	04	1.0	20.00	25.00	45.00	70.00
01/30/04	1908	PM	01	.5	10.00	20.00	30.00	100.00

FIGURE 3–3 A sample equipment history form

maintenance care than one whose failure does not have a direct or immediate impact on the plant operations, or one that has a backup that can take over in case of a failure.

- **Safety**. Equipment that can create a life-threatening situation or may cause serious damage to property if it fails needs a high level of maintenance.

- **History**. Equipment that is subject to regular breakdowns requires constant attention.

3.3.1 Three Degrees of Equipment Criticality

Three degrees of equipment criticality can be established to set the priority and level of preventive and predictive maintenance requirements. The system of criticality can be established with input from maintenance engineers, operators, and plant management.

Criticality Code I Equipment that has no backup and/or *will* cause bodily harm to people, damage other equipment, create a hazard to the environment, stop production, or create substantial financial loss if it fails falls in this classification. Production and material handling systems that will adversely and severely affect the operational capability of a plant if they break down may be given Criticality Code I. Certain components of hazardous materials conveyance systems, whose failure could be catastrophic and possibly could result in death, injury, chemical spills, and so on are also included in this category. Perhaps classifying the appropriate equipment as Criticality Code I may have averted the disastrous performance of the company responsible for generating electrical power for the city of Chicago and Cook County during the summer of 1999. (See the case study at the end of this chapter.)

The preventive maintenance program for Criticality Code I equipment should be designed to minimized breakdowns. Since the failure of such systems normally constitutes much greater financial loss, in the long run high levels of PM can prove to be economical and prudent.

Criticality Code II Equipment in this classification is still important and critical to production, and although its breakdown is undesirable, the results of a failure are not as drastic. The failure of such equipment may still create a loss or a slowdown in production, but the system can recover because the failure periods are short, or spare parts, substitute components, and backups are readily available for these pieces of equipment.

A normal preventive maintenance program should be designed to minimize the number of breakdowns and shorten the breakdown periods of Criticality Code II equipment while keeping maintenance costs in check.

Criticality Code III The breakdown or failure of this category of equipment will not seriously affect the normal operations of the facilities. The equipment is not frequently used, is sufficiently redundant, or its failure seldom affects production. Preventive maintenance activities for this type of equipment are minimal and are limited to normal cleaning, inspection, and adjustments for quality purposes. Anything beyond these basics should be clearly cost justified.

3.4 PLANNING FOR PREVENTIVE MAINTENANCE

As noted previously, the two main criteria used to plan an effective preventive maintenance program are equipment history and a system of criticality. Once the extent of preventive maintenance needed for a piece of equipment has been established, a PM plan should include the following six items:

1. A checklist that clearly indicates the specific preventive maintenance actions required, such as cleaning, inspection, and so on, and inspection frequency

2. Who will perform the PM, the operator or the maintenance department

3. Written standards for normal operating conditions for the equipment:
   - Normal operation temperature
   - Where to look for leaks and how much leakage is normal
   - Sounds that indicate the equipment is functioning properly or that there is a problem
   - Location of sounds
   - What is loose and how tight is too tight
   - Normal operating speeds and feeds
   - Signs of normal or abnormal wear
   - Lubrication points and the type and amount of lubrication to use
   - Sequence of inspection steps and what to look for during the inspection
   - Estimated required time

4. PM reports that show preventive maintenance steps taken, abnormalities detected, and corrective actions taken or to be scheduled

5. A PM route that reduces excessive traveling or backtracking by personnel

6. PM work order forms that clearly indicate what tools, equipment, material, parts, and skilled personnel are required for each task to avoid delays and inefficient use of the maintenance crew

It is generally agreed and understood that most preventive maintenance programs must have the full support of the management and should start small. Once the program has shown progress and has been accepted by the operating personnel, it can grow and become more comprehensive. The program often starts in one geographical area (one part of the plant) or as a small function (cleaning and housekeeping) and then gains acceptability and grows.

A successful program is one that attains the broadest acceptability. One secret to attaining this acceptability is to allow the participants to provide input when the preventive maintenance program is designed. Players with a stake in the preventive maintenance plan may include the following:

Equipment manufacturer. The manufacturer can provide detailed information regarding scheduling and frequency of preventive maintenance, type of lubricants to use, and all other equipment specifications, tolerances, and standards.

Maintenance department. The expertise and cooperation of maintenance staff is required because they can provide specific information about equipment maintenance and scheduling.

Operators. They have the closest relationship with the equipment on a daily basis. They can feel the vibration, hear the sound, and smell the odor *before* there is a vibration, a sound, or an odor that anybody else can detect. The operator can be your best ally or your worst enemy in designing and implementing a preventive maintenance program—the choice is yours.

Engineering department. Personnel in this department can help with rebuilding, modification, and overhaul of equipment. Their expertise is needed especially in most areas of predictive maintenance and equipment condition analysis.

And of course, in addition to input from personnel, information about the current equipment condition, current downtime and losses, safety issues, and historical data are needed to design and implement a successful PM program.

3.5   DESIGN FOR MAINTAINABILITY

A good preventive maintenance program does not start with the installation of the equipment nor does it begin at the time of equipment acquisition. Probably the most important aspect of a preventive maintenance program is conceived at the equipment design stage. Planning for preventive maintenance must be an integral part of the design stage.

Through proper design, the need and frequency of some maintenance activities can be drastically reduced and many functions can be simplified. Those functions that cannot be eliminated by design should be made simple to perform. The more difficult, time-consuming, complicated, or expensive the maintenance function, the more likely it will not be performed, it will be delayed, or it will be performed in a less effective or incomplete fashion.

Consider your personal vehicle. Are you more likely to perform a regular oil change if the manufacturer's suggested frequency is once every 3,000 miles or if it is once every 10,000 miles? How would the cost in performing the task or the degree of difficulty in reaching and replacing the oil filter and the drain nut affect the regularity with which you do this function? Even if you have your dealer perform these tasks, the ease or difficulty of each maintenance function is reflected in the labor cost.

Some maintenance functions can be totally eliminated by design. Not too long ago, car batteries had to have distilled water added in the cells regularly. Of course, quite often people would add regular water since distilled water was not readily available, causing long-term and adverse consequences as the result of minerals and impurities contained in tap water, or else would forget to check the battery all together and allow the cell to dry up, reducing the life of the battery. By redesigning and sealing the batteries, this particular maintenance function has been eliminated altogether. In so doing, not only has the maintenance function been eliminated, but all potential failures due to improper maintenance have been averted as well.

When it is necessary to perform routine functions, the design must include appropriate warning signals, preferably audio and visual, such as bells and lights, to warn the operator of needed action. Equipment should be designed to protect itself when required maintenance is not performed. A certain type of lawnmower stops running when the engine oil pressure drops below the safe operating level, *before* the engine is damaged. There are many examples of equipment designed with a built-in maintenance schedule, self-diagnosis, and the intelligence to call the service center when the need arises. The design and accessibility of the equipment must be simple and accommodating for maintenance purposes. Efforts must be made to design equipment to use standard tools and fasteners for various maintenance functions; basic maintenance routines should not entail the use of unique and special tools.

Engineers and managers responsible for the acquisition of new equipment must look for these kinds of features and insist on a design that incorporates the concepts of elimination, simplification, and standardization of maintenance steps. Design features that encourage and facilitate participative preventive maintenance on the part of the operators represent a wise investment in prolonging the productive life of the equipment.

3.6   COST OF PREVENTIVE MAINTENANCE

Management must see an acceptable return on investment (ROI) before they commit to or allocate additional investment for a preventive maintenance program. It can be argued that some basic level of preventive maintenance, such as cleaning and routine lubrication, is necessary for almost every piece of equipment. It may be impossible to separate the intrinsic return on investment of a few minutes of the operator's time per day from the tangible financial benefits. More extensive preventive maintenance activities, however, do require justification. For example, when a company concludes that the overall costs of repairs and loss of production caused by equipment breakdowns does not exceed the cost of PM, instituting a preventive maintenance program is questionable. If plant configuration, equipment redundancy, and safety factors are such that equipment failure does not impede production capabilities, a preventive maintenance program may prove to be unnecessary.

The following example illustrates the use of historical data in order to compare the cost of a preventive maintenance program with the cost of breakdown to select the least expensive alternative.

The following data indicate the breakdown frequency for specific pieces of equipment during the past two years. The history also indicates that each failure has cost the company $750 in lost time, repairs, and other failure-related expenses.

Number of failures	Number of months for failures
0	5
1	6
2	7
3	4
4	2
	Total months = 24

The summary table indicates the number of failures that have occurred in the past 24 months. How much should the company be willing to pay per month for a preventive maintenance program?

In order to answer this question, we need to calculate the *expected* number of breakdowns per month from historical data. Data indicates that there were five months when no failures occurred; in six months during the past two years, the company had only one failure per month; and finally, the company never had more than four failures during a given month.

The number of expected breakdowns is the weighted monthly average of breakdowns:

$$\text{Expected breakdowns} = \frac{\Sigma(\text{Number of failures}) \times (\text{Frequency})}{\text{Total frequency}}$$

$$\text{Expected breakdowns} = \frac{(0)(5) + (1)(6) + (2)(7) + (3)(4) + (4)(2)}{24}$$

$$\text{Expected breakdowns} = 40 / 24 = 1.67 \text{ failures per month}$$

The calculations show that the company has had a total of 40 breakdowns in the past 24 months, with a total cost of $30,000. The rate of expected breakdowns is approximately 1.67 breakdowns per month, with an expected cost (1.67 × $750) of nearly $1,252.50. If a preventive maintenance program with a lower price tag could be proposed that would virtually eliminate all breakdowns, then that would be a less expensive alternative.

The probability of a failure always exists, however, no matter how small. To continue with this example, let us assume that a preventive maintenance program costing $900 per month has been adopted and has been in effect for the past 12 months. Given the summary data below, do you recommend continuing with this PM program? Keep in mind that failures still do occur, and generate costs, in addition to the fixed cost of the PM program, $750 per occurrence.

Number of failures	Number of months for failures
0	10
1	2
	Total months = 12

Now with PM program in place, there have been 10 months with no equipment failure and two months with only one failure each.

$$\text{Total failures} = \Sigma (\text{Number of failures}) \times (\text{Frequency})$$

$$= (0) (10) + (1) (2)$$

$$\text{Total failures} = 2 \text{ for the past 12 months}$$

$$2 / 12 = 0.17 \text{ average failure per month}$$

The average monthly cost for breakdowns since instituting the preventive maintenance program is

$$0.17 \times \$750 = \$127.50$$

Total cost = Fixed PM cost + Current failure costs per month

Total cost = 900 + 127.50

Total cost = $1027.50

The current preventive maintenance program is saving the company $225 per month, which represents an 18% reduction in the monthly maintenance costs.

The preceding discussion is illustrated graphically in the graph in Figure 3–4, which presents a comparison between the cost of not having a preventive maintenance program and the cost of a preventive maintenance program as a function of the required or desired reliability level. The numerical data are presented in Table 3–2. The assumptions for the example are given at the top of Figure 3–4.

The horizontal line in the graph represents the cost of not having a maintenance program. This cost is obtained by multiplying the failure cost by the number of expected failures per year. The number of expected failures per year can be estimated by dividing the number of available hours per year by the MTBF. This ratio represents the expected number of failures per year when no preventive maintenance is performed.

Cost of having no PM = Cost per failure ×
Expected number of failures per year

Expected number of failures per year =
Available hours per year / MTBF

Expected number of failures per year = 2400 / 150 = 16

Cost of having no PM program = 6000 × 16 = $96,000

The preventive maintenance cost line is determined using various desired reliability levels. Using the mathematical expression for reliability, a preventive maintenance frequency (time between PM procedures) for a given reliability level is obtained. The number of annual PM procedures then can be calculated by dividing the available number of hours per year by the PM frequency. Part of the PM cost is then obtained by multiplying the number of PM procedures by the cost of each PM routine. Since for any given reliability level, a certain number of failures still can be expected, the cost of these failures is included in the total PM cost.

Failure Cost = $6,000.00
PM Cost = $120.00
MTBF = 150
Hours per Year = 2400

No Maintenance Cost = $96,000.00

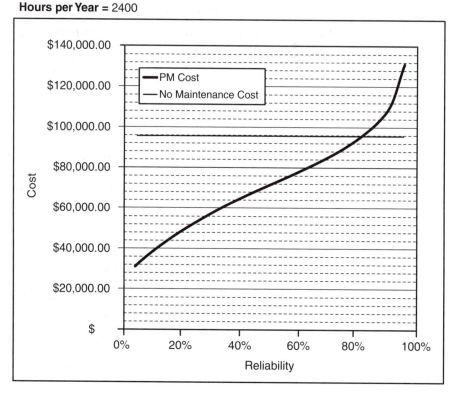

FIGURE 3–4   A graphical comparison of PM costs

This cost is calculated by multiplying the number of PM procedures by the failure risk (1 − Reliability) for the cost per failure:

Number of annual PM = Available hours per year / PM frequency

= 2400 / 8 (PM frequency for 5% failure risk) = 300

Annual PM cost = Number of annual PM × Cost per PM

= 300 × 120 = $36,000

Cost of failures (with PM) =

Number of PM per year × (1 − Reliability) × Cost per failure

= 300 × 0.05 × 6000 = $90,000

TABLE 3–2
Total PM Costs for Various Levels of Desired Reliability

Failure risk	PM frequency	PM cost	Reliability
5%	8	$131,011.28	95%
10%	16	$109,338.87	90%
15%	24	$100,419.07	85%
20%	33	$ 94,647.59	80%
25%	43	$ 90,099.46	75%
30%	54	$ 86,128.84	70%
35%	65	$ 82,454.52	65%
40%	77	$ 78,931.04	60%
45%	90	$ 75,472.08	55%
50%	104	$ 72,019.34	50%
55%	120	$ 68,527.83	45%
60%	137	$ 64,957.55	40%
65%	157	$ 61,267.52	35%
70%	181	$ 57,409.93	30%
75%	208	$ 53,322.01	25%
80%	241	$ 48,911.49	20%
85%	285	$ 44,024.63	15%
90%	345	$ 38,356.89	10%
95%	449	$ 31,084.22	5%

$$\text{Total cost of PM} = \text{Annual PM cost} + \text{Cost of failures (with PM)}$$
$$= 36000 + 90000 = \$126{,}000$$

Note that in Table 3–2, PM frequency hours are rounded to whole hours, for example, 8 hours. The computer program used to generate the graph and the PM cost, however, uses the actual frequency hours, such as 7.69 hours, so there is a discrepancy between the sample calculations presented in text here and the PM costs given in Table 3–2.

In general, the ROI for preventive maintenance programs is significant and often is manifested in increased productivity and improved quality, overall equipment effectiveness, better customer service, a more pleasant working environment, and improved morale.

As you examine the following case study, try to think *not* in terms of the costs of implementing a preventive maintenance program, but rather in terms of the costs of failure associated with the *lack* of an effective maintenance strategy. As illustrated in this case study, the costs associated with the lack of preventive maintenance programs often may prove to be too high to bear by most organizations.

THE PLIGHT OF COMED

Commonwealth Edison (ComEd), providing electrical power to over 3.4 million customers in northern Illinois including Chicago, experienced severe power outages that affected more than 100,000 residential and business customers during the summers of the late 1990s and 2000. In addition to significant economic and financial losses, these outages created severe health hazards and resulted in many deaths. During the extreme heat in the summer of 1995, ComEd's outages contributed to the deaths of hundreds of people.

Some of the business centers affected by these outages include the Chicago Loop, downtown office skyscrapers, the Chicago Board of Trade, and the Dirksen Federal Building, with extensive disruption of trade, business, and productivity.

"Every summer, it seems to be something different," noted an attorney for the Citizens' Utility Board. He continued, "Last summer it was power shortages. This summer it is problems with the transmission and distribution system." And an angry mayor, calling ComEd to task, declared that "They have major problems in that company. They have a responsibility to the city of Chicago and the city deserves answers."

What went wrong at such a giant corporation to cause so much "embarrassment" to the company, which resulted in the firing of several of its top executives and senior vice presidents?

A report to the Illinois Commerce Commission puts the blame clearly and squarely on ComEd's failure to spend money to *maintain* its substations and power lines. The report states that some of ComEd's circuits are more than 70 years old and in dire need of replacement. In addition, the company has failed to protect some of its substations against lightning strikes. The company planned for temperatures no higher than 93 degrees Fahrenheit, even though July temperatures routinely reach 96 degrees. These high temperatures stress the system and caused it to collapse repeatedly, leaving thousands of customers without power, sometimes for several days. Failure or breakdown of a substation or a transformer tends to overload other poorly maintained and already loaded transformers and substations, and so on, creating a domino effect of breakdowns and exacerbating the problem. Each subsequent failure affects a larger number of customers. The Citizens' Utility Board agreed that the report underscores concerns about ComEd's maintenance of its system.

A senior vice president of ComEd also said that investigation of the infrastructure had revealed that most of the system's problems were caused by a lack of maintenance. The ComEd chairman called the blackouts, "totally unacceptable" and implied that some employees could lose their jobs as a result of the outages. He added, "We must improve our maintenance and inspection procedures."

In summer 2000, as a public relations stunt, ComEd offered to pay $60 to residential customers and $100 to business customers if they suffered blackouts lasting a specified period of time. For the month of June 2000 alone, the company paid out $147,500. These are not the only losses that the company has endured. Lack of proper preventive and predictive maintenance procedures and programs, which have resulted in outages, have cost ComEd and its customers hundreds of thousands of dollars and also have resulted in the loss of lives.

In the light of (no pun intended) ComEd's troubles, Mayor Daley of Chicago introduced legislation allowing a Wisconsin company to build a power plant in Chicago. The mayor stated that this new plant provided much needed competition for the city's power monopoly. ComEd is also losing customers to other competitors. One of the biggest real estate management companies in Chicago recently defected from ComEd. Equity Office Properties Trust, the owner of 25 buildings totaling nearly 10 million square feet of space in the Chicago area, is now buying power from an alternate source. Among the other defectors are the owners of the Sears Tower and the John Hancock Center, two of the most notable Chicago landmarks and prominent features of the magnificent Chicago skyline.

Sources

Almer, E. "ComEd Outlines Repairs, Announces Resignation." *Chicago Tribune*, 1999. http://www.chicagotribune.com/news/metro/Chicago/article/0,2669,2-33820,FF.html.

LeBien, M. "ComEd Chief Offers No Excuses." *Chicago Tribune*, 1999. http://www.chicago-tribune.com/news/metro/Chicago/article/0,2669,2-32879,FF.html.

LeBien, M. "ComEd Vows to Restore Power by Friday Morning." *Chicago Tribune*, 1999. http://www.chicagotribune.com/news/metro/Chicago/article/0,2669,ART-32828,00.html.

Manor, R. "ComEd Says It's Improving." *Chicago Sun-Times*, 2000. http://www.suntimes.com/output/business/comed02.html.

Manor, R. "Report Faults ComEd's Maintenance Spending." *Chicago Sun-Times*, 2000. http://www.suntimes.com/output/business/comed0.html.

Roeder, D. "Zell Pulls the Plug on ComEd Power." *Chicago Sun-Times*, 2000. http://www.suntimes.com/output/roeder/col09.html.

QUESTIONS

1. Define preventive maintenance.
2. Why are economic considerations important in planning preventive maintenance activities?
3. Ideally, what percent of the total maintenance activities should be preventive?

4. What is meant by "empowerment" and what role does it play in TPM?

5. What are some of the basic PM activities and why are they important?

6. What is the importance of maintaining equipment history in order to have a viable PM program?

7. What is meant by a system of criticality and what do the various criticality codes mean?

8. In performing routine PM, why is the creation of a checklist important?

9. Planning for maintenance should begin at what stage of the product life? Explain.

10. What is meant by "design for maintainability"?

11. What are some of the preventive maintenance activities that you would recommend to ComEd (see the case study) or any other utility company?

12. The following data indicate the breakdown frequency for particular pieces of equipment during the past two years. The history also indicates that each failure has cost the company $1,050 in lost time, repairs, and other failure-related expenses.

Number of failures	Number of months for failures
0	4
3	6
5	6
8	5
9	3
Total months = 24	

The summary table indicates the number of failures that have occurred in the past 24 months. How much should the company be willing to pay per month for a preventive maintenance program?

4

PREDICTIVE MAINTENANCE

Overview

Objectives

At the completion of the chapter, students should be able to

- Distinguish the differences among corrective maintenance, preventive maintenance, and predictive maintenance.
- Understand various predictive maintenance techniques.
- Understand vibration analysis.
- Become familiar with common chemical analysis techniques.
- Define various methods of predictive maintenance methods and their applications.

4.1   INTRODUCTION

As industrial and manufacturing organizations strive to achieve world-class status, they will find it necessary to implement a certain level of predictive maintenance (PDM) technology into their operating procedures. Table 4–1 compares the existing and desired levels of predictive maintenance in various types of industries. The table suggests that on average, more than one-third of all maintenance activities in an industrial organization need to be of the predictive nature. Considering the current levels at which it is practiced, considerable strides still need to be taken to fully realize the benefits of predictive maintenance and its contributions to safety, resource utilization, and overall productivity.

4.1.1   Definition of Predictive Maintenance

Predictive maintenance applies various technologies and analytical tools to measure and monitor various system and component operating characteristics and to compare these data with established and known standards and specifications in order to predict (forecast) system or component failures. Whereas corrective maintenance is applied *after* the failure and preventive maintenance uses precautionary measures to avert possible problems, predictive maintenance actually evaluates the existing equipment condition and, based on a projected trend of the deterioration, failures are predicted and appropriate steps are taken. A simple example contrasting preventive and predictive maintenance is changing the oil in your vehicle. A preventive maintenance schedule may call for an oil change every three months or every 3,000 miles. This

TABLE 4–1

Current Versus Desired Levels of Predictive Maintenance in North America

Type of industry	Actual	World class
Assembly	7%	39%
Distribution	10%	30%
Manufacturing (Large)	12%	30%
Manufacturing (Small)	12%	32%
Process	15%	42%
Consultants' Opinion	15%	38%
Weighted Average	13%	35%

Source: Courtesy of Edward H. Hartmann, International TPM Institute, Inc.

prescriptive schedule does not bring into account an array of varying driving conditions. Predictive maintenance, on the other hand, analyzes the physical and chemical characteristics of the engine oil on regular intervals. Based on data analysis on the condition and properties of the engine oil, not the length of the service time, an oil change is scheduled. For this reason, predictive maintenance also is referred to as condition-based maintenance.

The operator or maintenance personnel, as part of the routine preventive maintenance procedures, automatically perform some basic predictive maintenance activities. A worn belt, a visibly vibrating shaft, or a misaligned bearing is quite obvious to a casual observer during a routine PM activity. Conclusions simply can be drawn that if these conditions are not corrected rather quickly, serious consequences are looming on the horizon. The mission and objectives of predictive maintenance, however, go beyond finding these superficial maladies to discover potential problems before the eye can see, the ear can hear, or the nose can smell them. To this end, predictive maintenance requires a greater investment in time, advanced technologies and equipment, and well-trained maintenance professionals, and therefore demands a greater commitment from management.

4.1.2 PDM's Reliance on Science and Technology

The field of predictive maintenance relies on various disciplines in science, engineering, and statistics to determine the current equipment conditions, compare these conditions with the established and acceptable degree of variation from the normal operating standards, and signal any abnormal conditions or trend. Chemical analysis, infrared scanning, ultrasonic imaging, vibration analysis and shock pulse measurements, ferrographic particle analysis, X rays, and various nondestructive testing methods are a few of the predictive maintenance tools available. Some of these and various other techniques will be discussed in this chapter.

4.1.3 Decision Factors Regarding PDM

Decision factors regarding implementation of PDM are quite similar to those that were discussed previously for preventive maintenance planning. Equipment criticality and economic justifications are the major considerations. Critical equipment, either in terms of safety, production losses, or the cost of the equipment itself, are prime candidates for predictive maintenance activities. The cost factor is a more serious consideration with predictive maintenance technologies than it may be with preventive techniques. Initial investment in measurement and monitoring equipment as well as the cost of training and skill levels of the personnel are significant factors. The initial investments and the continued costs of the operation must be weighed against the potential cost of failures and breakdowns. Therefore, a cost-benefit analysis and a clear statement of objectives for the return-on-investments (ROI) are important parts of PDM planning and implementation.

4.1.4   PDM's Quantitative Nature

Predictive maintenance is *quantitative* in nature. Data are collected, analyzed, charted, and interpreted. Based on these data, vital decisions are made. At the advent of the computer and information age, the cliché "garbage in, garbage out" was added to our American lexicon. Not only has the phrase not lost its applicability, given our increasing dependence on information, it is more relevant today than ever before. As it is the case with any analytical process, good, reliable, and relevant data are priceless, but inaccurate and faulty data can be extremely expensive.

It is important to understand the purpose for any data collection because it can be time-consuming and therefore costly. Do not collect data if you do not intend to use it. Maintain the integrity of the data. Make sure you know how, when, where, and by whom the data were collected. The conditions under which the data were collected will have a direct bearing on the findings. The measurement instrument itself can cause error, create bias, and lead to faulty conclusions. Therefore, an adequate calibration and maintenance plan for the measurement instruments becomes an important issue in itself.

4.1.5   Approaches to Data Collection

The two general data collection methods are fixed and portable. Fixed-type devices are remote collection systems most appropriate for harsh environments that are not readily accessible to human agents. These devices may be installed permanently to monitor equipment conditions and are capable of collecting and transmitting the data. Portable-type devices can be taken from one piece of equipment to another for the purpose of data collection. Automated data collection and recording systems, such as radio-frequency devices, sensors, and various microprocessors, tend to reduce the human error factor.

4.1.6   Data Analysis

The data that have been collected must be analyzed. The analysis may be chemical, engineering, statistical, or very likely a combination of these. The analysis of an oil sample or cutting fluids can determine the degree of molecular breakdown and levels of contamination, for example. In the case of vibration analysis, the analysis may require comparison of vibration levels for a specific piece of equipment with known vibration limits. The outcome of the analysis then must be compared with established standards or a "baseline." The equipment manufacturer often provides these limits or baselines.

Data analysis and trends lead to conclusions. Trend lines and regression models, numerical or graphical, can forecast the future behavior of the equipment. Past or historical data can be projected to predict future trends. However, do not attempt to generalize beyond the scope of the data you have collected and be aware that not all trend lines tend to be linear. Data collected for a particular class of equipment under a particular set of conditions can lead to erroneous conclusions regarding the future pattern of behavior for a different class of equipment or under a different set of conditions. The history of the data is as important as the data itself.

Once the equipment status has been determined, decisions on the course of action are taken. The corrective actions and the time frame for the actions are determined and executed. Of course, the decision may be to take no action. If the cost of corrective steps exceeds the cost of failure, it may be feasible to do nothing.

4.2 PREDICTIVE MAINTENANCE TECHNIQUES

In the following sections, some basic and common predictive maintenance techniques, equipment, and applications are introduced and briefly discussed.

4.2.1 Vibration Analysis

Most operating equipment experiences some level of vibration. Ideally the vibration levels for all equipment with rotating parts should be regularly monitored, but this option may not be feasible based on economic and staffing considerations.

Vibration is an excellent early warning signal pointing to equipment deterioration. The common causes of vibration are imbalance of a rotating part, misalignment, defective bearings, and defective belts. In most cases, if vibration can be observed by human senses such as touch, sight, or hearing, it is probably already too late and some damage has taken place. Vibration monitoring can detect these early warning signals and prevent equipment failure. In most cases, especially with mechanical components in which friction is minimal, vibration monitoring is an effective predictive maintenance strategy and can provide earlier warning signals than temperature monitoring.

Factors affecting vibration levels Practically every piece of equipment experiences some level of vibration. Different types of equipment under various operating conditions experience different levels of vibration. A number of factors affect the observed vibration level. These factors include the size, stiffness, and weight of the equipment, the rigidity of the base on which it is mounted, and the surrounding equipment. Other pieces of operating equipment, connecting pipes, and ductwork can add to the vibrations generated internally and can be just as destructive to the machine. Certain types of equipment, such as pumps, compressors, and reciprocating equipment in general, normally exhibit higher levels of *acceptable* vibration than rotating equipment; therefore, allowances must be made when determining equipment condition. Variation in the measurement instruments and their behavior at various vibration levels can and will introduce discrepancy in the readings. It is important to use the same instrument regularly for monitoring the same equipment in order to reduce error in equipment monitoring. Also, since various measurement points give different vibration readings, it is important to keep the monitoring points constant. Always monitor the equipment at its normal operating conditions, such as temperature and load. If the equipment is operated under various loads or speeds, it should be monitored at all conditions and the measurements should be collected and charted separately. The level of criticality often determines monitoring frequency. In general, most machines should be monitored monthly.

Vibration standards ISO Standard 2372 sets some general standards and guidelines for vibration monitoring based on equipment classification and size. Table 4–2 shows four equipment classifications, various levels of vibration severity, and the extent to which the vibration level is acceptable for a given class of equipment. Vibration velocity (inches or millimeters per second) of rotating equipment in effect is a measure of forces on the bearings. Vibration severity of a machine is defined as the maximum root-mean-square of the vibration velocity. Equipment classification is based on the following criteria:

Class I This class includes small machines and individual parts of engines and machines integrally connected with the complete machine in its normal operating condition. Electrical motors up to 15kW are examples of machines in this category.

Class II Medium-size machines, such as electrical motors with 15kW to 75kW output, without special foundations, fall in this category. Also, rigidly mounted machines up to 300kW on special foundations are included in this class.

TABLE 4–2

Velocity Range Limits and Machinery Classes—ISO Standard 2372

Vibration Severity		Velocity Range Limits and Machinery Classes: ISO STANDARD 2372				
		Small machines	Medium machines	Large machines		Severity
Vrms/mm/s	V PEAK/ips	Class I	Class II	Rigid supports Class III	Flexible supports Class IV	
0.28	0.02		GOOD			A
0.45	0.03					
0.71	0.04					
1.12	0.06					
1.80	0.10					
2.80	0.16		SATISFACTORY			B
4.50	0.25					
7.10	0.39		UNSATISFACTORY			
11.2	0.62					C
18.0	1.00					
28.0	1.56		UNACCEPTABLE			D
45.0	2.50					
71.0	3.95					

Source: Courtesy of SKF Reliability Systems.

Class III In this category are large prime movers and other large machines with rotating masses mounted on rigid and heavy foundations that are relatively stiff in the direction of vibration measurement.

Class IV Large prime movers and other large machines with rotating masses mounted on foundations that are relatively soft in the direction of vibration measurement are in this class. Examples are turbogenerator sets, especially those with lightweight substructures.

Vibration has an adverse effect on a bearing's life because bearings absorb the harmful vibration energy. Small amount of vibration can reduce the useful life of a bearing significantly and, if undetected, can result in untimely failures. Vibration monitoring can detect unacceptable levels of variation and potentially avert costly failures. Studies show that vibration analysis, as part of a predictive maintenance program, results in significant returns on the investments of time and equipment. Table 4–3 shows the effects of various levels of vibration of equipment condition and subsequent consequences if the condition is left untreated. Excessive vibration is not only harmful to the equipment but also can damage the surrounding equipment, walls, and building. Vibration and

TABLE 4–3

Effect of Vibration on Equipment Condition

Vibration velocity	Equipment condition
0.15 ips (or less)	Low force level; bearing life is 10 to 16 years minimum with proper lubrication.
0.30 ips	Double the normal force level; bearing life is decreased by a factor of 8, or $1^1/2$ to 2 years with proper lubrication.
0.60 ips	Very high forces; bearing life is now only 6 to 8 weeks. Force level is high enough to rupture the surface tension of an oil film and make lubrication ineffective.
0.90 ips	Extremely high forces; bearing is damaged with every revolution. Bearing life ranges from 3 days to a few weeks.

Source: Courtesy of Edward H. Hartmann, International TPM Institute, Inc.

the resulting noise levels in excess of OSHA standards are harmful to the operation personnel as well and can cause fatigue, blurred vision, loss of balance, and other adverse effects.

The main objectives of vibration monitoring and analyses are to measure the changes in amplitude of variation by frequency over time. The amplitude is plotted on the x-y coordinate system and is tracked over time for a given load. This is referred to as the *vibration signature*. Changes to the vibration signature of equipment signal a change in the characteristics of one of the rotating element, such as a bearing, shaft, or other components.

Periodically, Wade Utility, located in West Lafayette, Indiana, employs the services of the Machinery Health Monitoring Company of Indianapolis to collect vibration data from various equipment and obtain an analysis of any possible problems. The top portion of Figure 4–1 provides the graph of such data points collected over a period of time. The data points state the level of the equipment vibration and are plotted as inches per second (ips) over time. As shown in the top graph of Figure 4–1, the area between 0.314 ips and 0.628 ips, highlighted by the lightly shaded area, is referred to as the first alarm region. The heavily shaded area of the graph, from 0.628 ips and beyond, indicates the critical region.

The chart in Figure 4–2 is used to determine the degree of operating roughness and alarms. In order to utilize the roughness chart, the current equipment vibration reading from the trend analysis graph is transferred

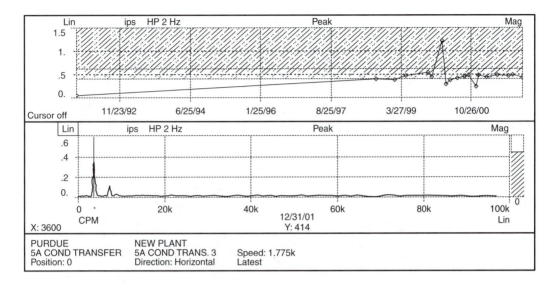

FIGURE 4–1   Vibration trend and signature
(Source: Courtesy of Wade Utility Plant.)

from the graph in Figure 4–1. In this case the last reading on the trend analysis (the top portion of Figure 4–1) is approximately 0.5 ips. Observe the diagonal lines on the chart in Figure 4–2. These lines indicate the degree of the roughness or the operating condition of the equipment. In this example, observe that the vibration reading of 0.5 falls between the diagonal lines of 0.314 ips and 0.628 ips, which indicates that the equipment is operating under rough to very rough conditions. Failure would appear to be imminent; however, the bearing has been operating at this level for more than a year. Immediate action may not seem necessary, but it is important to realize that the life of the bearing is severely threatened. Planned shutdown in the near future is necessary to replace the part and to perform a root cause analysis to avert a possible catastrophe.

The top portion of Figure 4–3 is the trend analysis from Wade Utility's boiler feed pump and shows a relatively stable trend over two years. The last vibration reading on the chart, taken shortly before October 8, 2001, indicates a reading of 0.168 ips. Once again, in order to determine the roughness level, the reading is transferred to the chart in Figure 4–2, where we observe that the vibration level of 0.168 falls between the two diagonal line values of 0.157 and 0.314, indicating a "slightly rough" operation condition. It is interesting to note that this machine has been running at the same vibration velocity for almost two years and has not yet crossed into the rough stage, which is also the first warning zone.

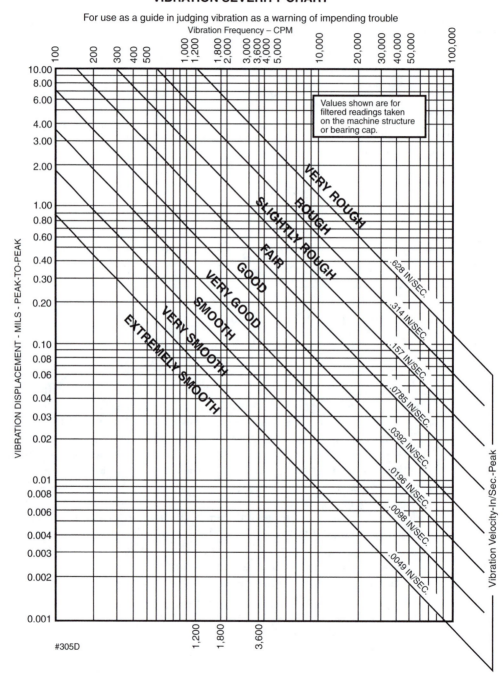

FIGURE 4–2 Vibration/roughness chart
(Source: Courtesy of Wade Utility Plant.)

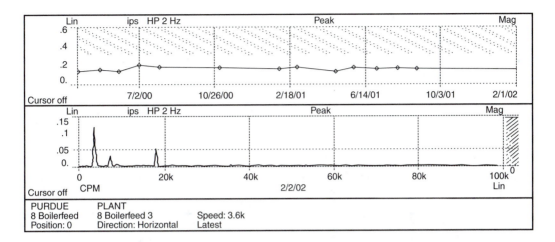

FIGURE 4–3 Vibration trend and signature for the boiler feed pump
(Source: Courtesy of Wade Utility Plant.)

The graph in the top portion of Figure 4–4 shows a trend analysis for the pump in Wade's E Cell tower. Review of the graph reveals that the most recent test resulted in a vibration reading of 0.222 ips. Again, by examining the roughness graph in Figure 4–2, we can determine that this vibration velocity is between 0.157 ips and 0.354 ips, which corresponds to the slightly rough stage. Although this is the same roughness level as that for the boiler feed pump (Figure 4–3), a closer examination of the graph in Figure 4–4 shows that the vibration velocity does seem to be increasing and that may be cause for alarm. It is constructive to study the graph carefully. The trend line clearly indicates that other jumps in the vibration velocity also occurred in the previous winter months. After further investigation, we discovered that this pump is located outside the plant, and during the colder winter months the pump has to operate at a faster speed to perform its function. It is expected that the vibration velocity will decrease once the climate turns warmer and the operation speed is returned to the normal level.

So far we have examined only the top portion of each of the graphs presented in Figures 4–1, 4–3, and 4–4. Each trend analysis graph in these figures has an accompanying vibration signature for the given piece of equipment, which is displayed in the lower portion of each of these figures.

Every piece of operating equipment exhibits a certain characteristic level of vibration, no matter how excellent or stable the operating conditions may be. These inherent vibrations, caused by various components and the general operational characteristics of the machine, generate a vibration "baseline" that is called the *vibration signature*. It is against this

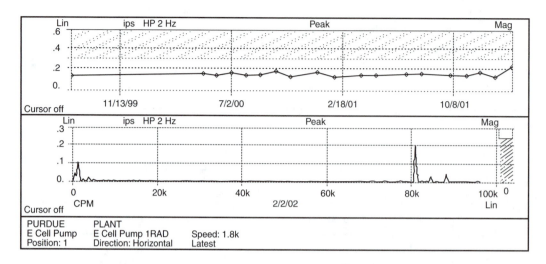

FIGURE 4–4 Vibration trend and signature for the E Cell pump
(Source: Courtesy of Wade Utility Plant.)

baseline or signature that any abnormality or out-of-control conditions can be detected and assessed. The vibration signature of a piece of equipment shows the overall vibration velocity for all of the components of that piece of equipment, or it can be thought of as the composite of the vibration of each individual component that makes up the equipment. For example, a fan motor may operate at 150 cycles per minute, the blades may run at 1,050 cycles per minute, and the drive shaft may operate at 1,800 cycles per minute. The overall or composite vibration velocity of these individual components will be represented along the y-axis. The operating velocity, given in cycles per minute, can be read on the x-axis. Each peak in the vibration signature theoretically can be attributed to a specific component in the equipment. However, the ability to decompose the signature to identify each individual component's contribution to the overall vibration level, that is, knowing which component is represented at which cycle per minute, takes in-depth knowledge of the equipment and requires additional research.

Note that simply because one component measures a velocity of 0.05 ips and another may measure 0.05 ips, or 10 times the vibration level of the first component, it does not necessarily mean that the component producing the measurement of 0.5 ips is the problem component. The component with 0.05 ips velocity may be much more critical than the second. Knowledge of the equipment and its components is crucial in making this determination.

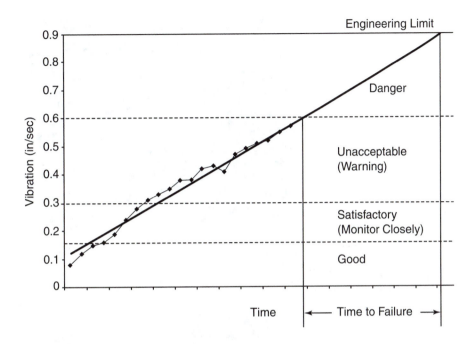

FIGURE 4–5   Vibration signature and engineering limits

Vibration monitoring equipment   A variety of monitoring devices, portable or built in, are used to collect and graph the data and develop the vibration signature. Data can be trended and used to forecast the expected time to failure. Figure 4–5 displays a hypothetical vibration signature for a machine and the expected failure based on the engineering limits and the regression line. Figures 4–6 through 4–9 display various vibration monitoring devices.

Vibration sensors   Selecting the appropriate type of vibration transducer or sensor is an important step when implementing a vibration analysis program. A decision about the type of transducer to select should be based on the application in which it will be used.

There are three different types of transducers on the market. These are acceleration, displacement, and velocity. Acceleration sensors, or accelerometers, are used for higher frequencies or when force is analyzed. Displacement sensors are used for lower frequencies and when displacement amplitude is important. Velocity sensors are used for intermediate frequencies and when displacements are small.

Recall from physics that velocity is defined as the time rate of change of displacement; acceleration is the time rate of change of velocity.

FIGURE 4–6   Vibration transducers
(Source: Courtesy of CEC Vibration Products.)

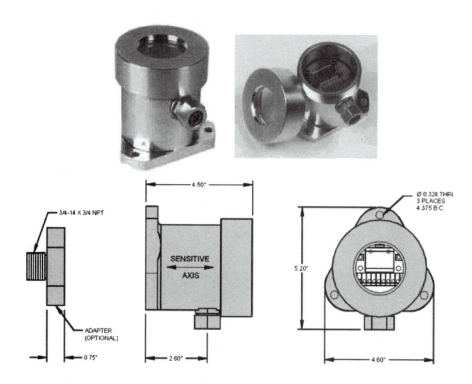

FIGURE 4–7   A vibration switch
(Source: Courtesy of CEC Vibration Products.)

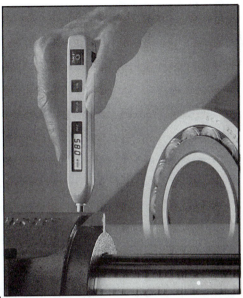

FIGURE 4–8   A compact, lightweight instrument for bearing fault detection
(Source: Courtesy of SKF Reliability Systems.)

Vibration charts   For the purposes of vibration analysis, data are collected and charted in two main types of graphs that then can be analyzed: time waveforms and spectrums.

Time waveforms are constructed by plotting displacement as a function of time. Figure 4–10 displays a simple harmonic oscillation, the simplest form of a time waveform.

A Powerful Combination For Detection Of Machine & Bearing Defects

SKF Vibration Pen   SKF *SEE*© Pen CMBP30   SKF ThermoPen TMTP1

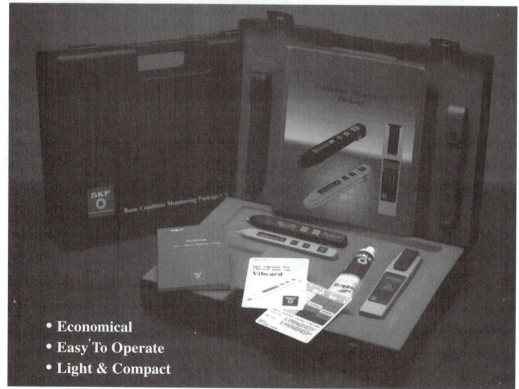

- **Economical**
- **Easy To Operate**
- **Light & Compact**

FIGURE 4–9 Versatility and ease of use of a lightweight vibration monitoring device

(Source: Courtesy of SKF Reliability Systems.)

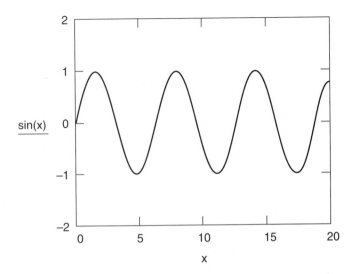

FIGURE 4–10   A simple harmonic oscillation

At any given time, however, several different factors and forces may act on the object, each producing its own oscillation at its own unique frequency. The resulting time waveform no longer resembles the simple harmonic wave that was observed in Figure 4–10; rather, it will be a composite of all the oscillations.

For simplicity, assume that there are only two vibration factors acting on a bearing at the same time but at different frequencies. Figure 4–11 illustrates the resulting time waveforms. Parts (a) and (b) of Figure 4–11 show the harmonic oscillation resulting from each of the factors separately. Part (c) of Figure 4–11 illustrates the composite or the overall effect of these two vibration factors on the part and the resulting time waveform.

In reality, many more than two vibration forces, each with its own characteristics and frequency, simultaneously act on the part. Although each vibration produces its own unique harmonic oscillation in isolation, the overall composite time waveform may appear as shown in Figure 4–12.

Given the complexity of the composite time waveform and the difficulty associated with its analysis and interpretation, it is often necessary to isolate the various time waveforms so that we can pinpoint more precisely the vibration forces acting on a part and then analyze them. In order to accomplish this, we use the time waveform data and create a spectrum of the waveform by separating the different frequencies. Figure 4–13 displays an example of such a spectrum. The resulting spectrum is

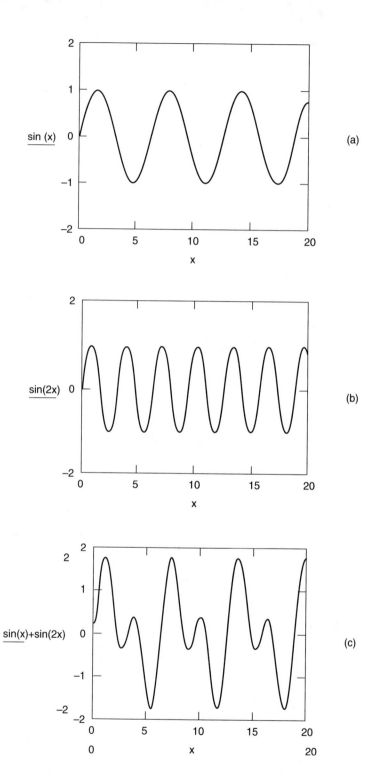

FIGURE 4–11   Effects of two forces on a part

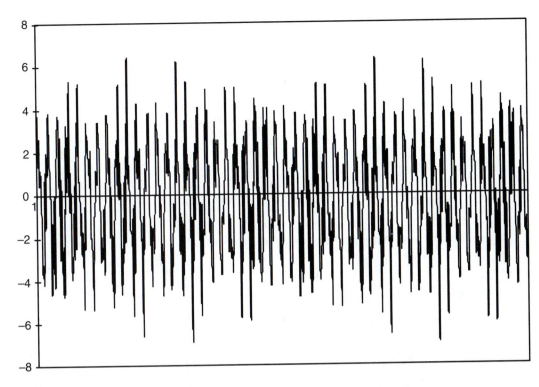

FIGURE 4–12   A time waveform resulting from many forces on a part

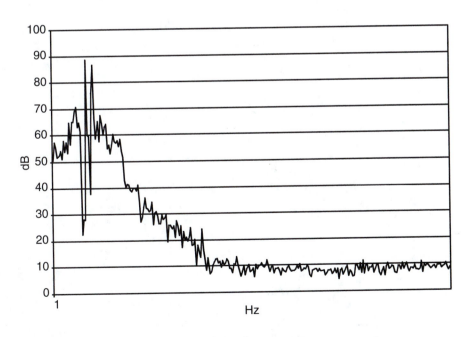

FIGURE 4–13   A spectrum of the forces acting on a part

much easier to analyze and interpret. More in-depth details about vibration analysis are beyond the scope of this text. Interested readers are referred to a wealth of knowledge and publications available on the subject.

The following case study illustrates the use of vibration monitoring equipment. The case clearly shows how seemingly minor vibrations can propagate quite aggressively and lead to disastrous outcomes if left undetected at early stages. The case also highlights the importance of a proactive approach to detecting and solving the problems of equipment vibration in order to avert major production and financial losses.

VIBRATION MONITORING AVERTS PLANT FAILURE

Iggesund Paperboard, a British manufacturer of coated carton board used by the packaging and pharmaceutical industries, was able to avert a catastrophic plant failure by investing in condition monitoring equipment and predictive maintenance practices. The company's use of condition monitoring equipment and associated software produced by SKF Engineering Products, Ltd. proved invaluable in providing advance warning of the imminent failure of a critical bearing on machinery at its manufacturing plant.

The timely warning and detection led to a planned repair during a scheduled shutdown that resulted in huge savings in unscheduled downtime and production losses. Had the bearing failed completely, the resulting damage to the plant would have been catastrophic and would have required shutting down the machine for six months.

As a part of their continuous monitoring program, the technicians detected a developing problem with the machine glaze (MG) cylinder bearing during a routine inspection. To appreciate the magnitude of the problem, it is helpful to understand and visualize the immense size of the equipment. The machine is more than 660 feet long and the cylinder has a diameter greater than 20 feet and weighs approximately 165 tons. The gigantic size of the machinery gives some indication of the severity of the problem had the failing bearing escaped detection.

Iggesund has progressively introduced condition monitoring and predictive maintenance practices into the plant. Figure 4–14 displays the results of one such vibration monitoring and data collection exercise. The measurements are collected over a four- to five-year period and display the vibration trend of the bearing cap on the MG cylinder. The baseline was established, showing a steady trend with a mean value of 0.03 ips rms. A sudden increase in the vibration level appears on the graph at the 0.05 ips rms level. Given that many machines on the Iggesund site typically operate at high vibration levels, this observed value might have been accepted if the baseline for the MG cylinder had not been established previously. However, since the MG cylinder is a slow-running machine, operating at an average speed of 12 rpm, and based on its established vibration baseline, a reading of 0.05 was considered high enough to warrant further investigation.

Further studies, including spectral analysis, showed that there was harmonic amplitude with a spacing of approximately 3 Hz, as shown in Figure 4–15. Again, this is very low, since defective bearings exhibit much higher levels of vibration with more clearly defined peaks. Because of the huge mass of the cylinder and its slow speed, however, the condition monitoring team decided to take a closer look at the problem. By identifying the roller as a double-row spherical roller, the team

was able to calculate the bearing defect frequencies. Based on the vibration spectrum characteristics, the team determined that the peaks indicated a bearing defect frequency. Although all bearings are prone to some slippage and sliding, in this case it was essential to determine the exact cause and source of the defect. A method known as acceleration enveloping was used to take a spectrum. Enveloping technique is used to enhance small signals by first separating higher frequency-bearing signals from low-frequency machine vibrations with the aid of a filter. The envelope approximately squares the defect signals. These signals, as shown in Figure 4–16, are repetitive and can be simulated and studied easily. Figure 4–17 exhibits a "zoomed" section of the spectrum. With the aid of the spectrum analysis, the

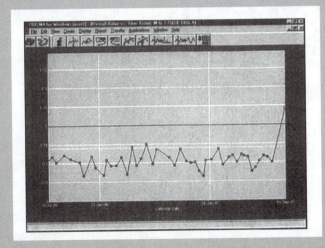

FIGURE 4–14   The overall vibration trend, showing a significant jump at the tail end

(Source: Courtesy of SKF Reliability Systems.)

(continued on next page)

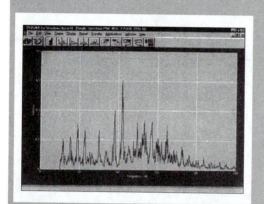

FIGURE 4–15 The vibration spectrum shows harmonic peaks
(Source: Courtesy of SKF Reliability Systems.)

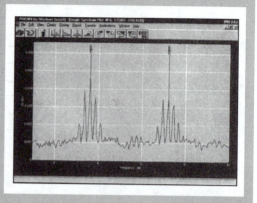

FIGURE 4–17 A "zoomed" spectrum shows bearing defect frequencies
(Source: Courtesy of SKF Reliability Systems.)

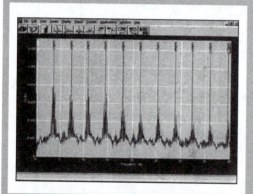

FIGURE 4–16 Enhancing small signals with the aid of "enveloping" technique
(Source: Courtesy of SKF Reliability Systems.)

team was able to positively identify the problem as an inner-race defect, most likely a raceway crack. The team continued to monitor the bearing closely on a daily basis. The trend indicated above normal amplitude, which tended to vary with the speed of the machine.

The defective bearing was eventually replaced during a *planned* shutdown. It was found that the inner bearing raceway of the defective bearing had two noticeable cracks. Damage to rotating equipment can be caused by many factors. Cracks in bearings are usually considered secondary damage resulting from primary factors such as wear and surface distress.

SKF personnel believe that the Iggesund condition monitoring team detected the second crack, which was caused by the stresses imposed by the original crack. Measurements taken on the bearing after replacement showed a major improvement in vibration levels, as shown in Figure 4–18.

data storage only not comparison.

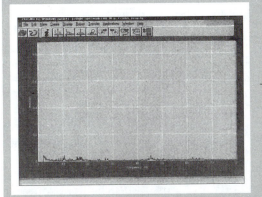

FIGURE 4–18 The vibration spectrum after the mini-shutdown for preventive maintenance
(Source: Courtesy of SKF Reliability Systems.)

Although the team at Iggesund successfully detected the inner-race defect in the MG cylinder and therefore averted a major catastrophe, procedures have been introduced to enable earlier detection of this type of problem. Iggesund plans to use a fully automated condition monitoring system. The system uses a dedicated computer linkup to continuously monitor all crucial rotating components throughout what is described as an extremely complex manufacturing process. Data will be collected and automatically entered into each component's profile. This will provide an instant comparison between the current measurements and previously recorded data. The information is stored for historical analysis of the machine, showing at a glance any deviation from normal performance.

Source

SKF Engineering Products, Ltd. *Revolutions*, vol. 7, no. 1.

4.2.2 Chemical Analysis

Chemical analysis is a method of predictive maintenance that allows us to study the internal conditions of the equipment, determine the cause of problems, predict any impending failures, and decide on a course of action. Just as chemical analysis of human body fluids in a medical laboratory, such as an analysis of blood chemistry, can give a clear indication of how various organs are working and the overall health of an individual, the condition of various fluids obtained from equipment can reveal the state of the internal health of the machine. Coolants and lubricants are the lifeblood of equipment and can reveal a great deal of information about its internal condition. Analytical data show the level of deterioration and the type of contaminants in the lubricants, which point to various causes and abnormalities. The purpose of the analysis is not simply to decide when to replace the lubricants or add more oil, but to determine the underlying causes of deterioration and contamination and plan for appropriate remedy.

Common chemical analysis techniques Some chemical analysis techniques that are commonly used in predictive maintenance are introduced and briefly discussed in the following sections.

Spectrographic analysis This technique refers to any number of qualitative or quantitative analytical procedures involving any one of various types of spectroscopy to determine the presence and quantity of various elements or compounds in a sample. The method is based on the fact that when light energy is passed through matter, the matter absorbs a certain amount of this energy, and a certain portion is emitted. The amount of absorption and emission is characteristically unique for each substance and is analogous to identifying individuals using their fingerprints.

A spectrometer is used for spectrographic analysis, and a simple schematic of the instrument is shown in Figure 4–19. The light source, a source of radiation appropriate for the type of analysis, such as a light bulb, a gamma-ray emitter, or a laser, provides the source of energy. A sample of the compound, such as oil or coolant, is placed in the chamber, where it is vaporized or sprayed into flames to decompose it, freeing the atoms for absorption of the light energy. In the third stage of the analysis, all the wavelengths of radiation that are present and have not been absorbed by the sample are separated so that they can be identified and measured. The final stage of the analysis detects the amount of radiation that is emitted (the portion that is present) or has been absorbed (the portion that has been absorbed by the atoms). The presence of various elements in the sample can be determined by identifying the various wavelengths. This is the qualitative aspect of the analysis. The amount of radiation gives information regarding the quantity of the substance in the sample. As stated earlier, each element emits or absorbs its own characteristic wavelength and produces its own unique *spectrum*.

Most spectrometers designed for analysis of lubricants are equipped with a radiation source that is specifically designed for the detection and analysis of metal particles from the machinery that may be suspended in these fluids. In order to perform such an analysis, a small sample of lubricating fluid is taken from the equipment and placed in the vaporization chamber. The light emitted produces wavelengths characteristic of those

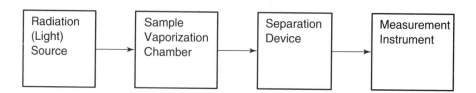

FIGURE 4–19 Schematic diagram of a basic spectrometer

elements that are present in the fluids. The intensity of each wavelength is measured to determine the concentration or the quantitative value of the particles in the sample. Through the use of this analytical technique, it is possible to detect abnormal wear and other malfunctions far in advance of the time at which they would become serious problems.

The oil vendors and suppliers of various lubricants can provide the baseline spectra, which indicate the normal or expected chemistry of their products. These baselines then can be used as the basis of comparison with the results of periodic analyses to determine the change in the internal condition of equipment as indicated by the changes in the chemical analysis. Spectrometry or spectrochemical analysis can be used to detect the presence of numerous elements, airborne pollutants, and other contaminants, each of which can point to a possible problem requiring a solution. Table 4–4 shows a partial listing of elements that may be detected through the use of spectroscopy and some of the possible sources of these contaminants.

The presence of these and other particles in the lubricants are both the effect and the cause of further equipment problems. The presence of

TABLE 4–4

A Partial List of Elements Found in Lubricant and Possible Sources

Element	Source
Iron (Fe)	Cylinders, gears, pistons, crankshafts, bearings, housings, rust
Chromium (Cr)	Rings, bearings, plating
Copper (Cu)	Bushings, bearings, washers, friction plates
Tin (Sn)	Bearings, bushings, pistons
Aluminum (Al)	Pistons, pumps, bearings, rotors, blowers
Nickel (Ni)	Valves
Silver (Ag)	Plating, bearings, bushings
Manganese (Mn)	Liners, rings
Silicon (Si)	Airborne dirt
Sodium (Na) Magnesium (Mn) Calcium (Ca) Phosphorus (P) Zinc (Zn)	Antifreeze and other additives

metal particles from wear signals abnormal equipment conditions such as improper alignment, components that are out of balance, inadequate clearances and tolerances, and so on, which contribute to excessive wear. The particulates in turn compound the problem by acting as abrasive agents and creating additional friction and wear.

pH monitoring　pH monitoring is used in a variety of applications including industrial, chemical, and pharmaceutical processes, as well as in environmental processes such as wastewater treatment.

The pH of a solution is measured to determine its level of acidity or alkalinity (base). More technically, the pH of a solution is a numerical value relating to the concentration of the hydrogen ions in a solution and the level of activity of these ions. This numerical value, the pH of a solution, can range from 0 to 14; a value of 0 indicates the highest level of acidity and 14 indicates the highest level of alkalinity of a solution. The value of 7 on this scale indicates a neutral solution such as pure water.

pH monitoring can play an important role in preventive and predictive maintenance practices by providing an insight into the internal condition of the equipment. Internal fluids of the machinery such as oils, coolants, lubricants, and so on have a prescribed range of acceptable pH as specified by the equipment manufacturer and the fluid vendors. pH monitoring allows us to detect any deviations from these specifications, which commonly point to some internal abnormality such as the presence of external contaminants caused by a bad seal, the breakdown of lubricants, or a host of other problems. Monitoring and tracking the pH of these fluids, which are the lifeblood of the operating equipment, can give advance warning of an impending equipment failure that often can be easily averted. Such monitoring is an important part of a complete preventive and predictive maintenance programs in industrial processes such as refining, power generation, semiconductor manufacturing, and metal processing, to name a few. Furthermore, as part of a strong preventive and predictive maintenance strategy, companies can evaluate and compare additive concentrations and make more informed and educated decisions.

Chemical analysis of lubricants can be carried out rather inexpensively. Distributors and local laboratories can provide these services at reasonable costs. For situations in which the expense of purchasing pH monitoring equipment can be justified, many types of instruments are available for in-house use. Figures 4–20 through 4–22 show examples of some of these portable and easy-to-use systems.

4.2.3　Tribology

Today the science of tribology has paved the way for a clear understanding of the effect of friction on material waste and energy dissipation. A

FIGURE 4–20   Examples of handheld chemical condition monitoring devices
(Source: Courtesy of Fluitec International.)

relatively new discipline, tribology, from the Greek work *tribos*, is the science and technology of friction, lubrication, and wear. Understanding and controlling friction and therefore material and energy conservation are important aspects of predictive and preventive maintenance. Of course the effect of friction on materials was recognized long before the formulation of the science of tribology. Leonardo da Vinci (1452–1519), the great Italian master of art, engineering, and science, whose scientific studies in hydraulics, optics, and anatomy have had a great influence on modern scientific advances, was concerned about the effect of friction on heavenly bodies. He wrote, ". . . had this friction really existed, in the

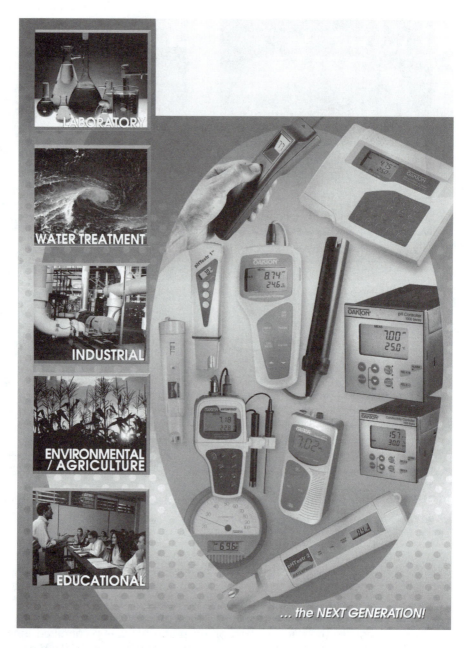

FIGURE 4–21   A vast array of portable and versatile pH monitoring devices
(Source: Courtesy of Oakton Instrument.)

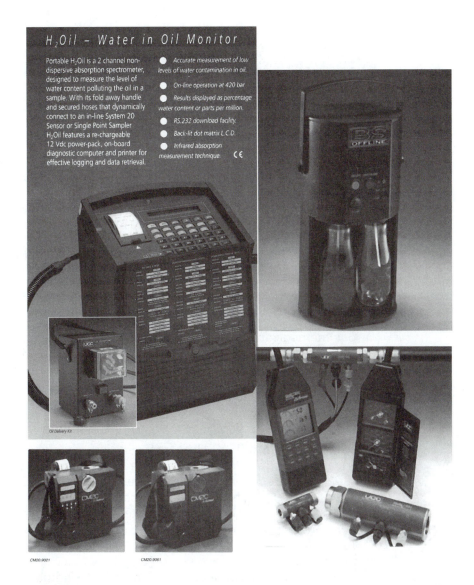

H₂Oil – Water in Oil Monitor

Portable H₂Oil is a 2 channel non-dispersive absorption spectrometer, designed to measure the level of water content polluting the oil in a sample. With its fold away handle and secured hoses that dynamically connect to an in-line System 20 Sensor or Single Point Sampler H₂Oil features a re-chargeable 12 Vdc power-pack, on-board diagnostic computer and printer for effective logging and data retrieval.

● Accurate measurement of low levels of water contamination in oil.

● On-line operation at 420 bar

● Results displayed as percentage water content or parts per million.

● RS.232 download facility.

● Back-lit dot matrix L.C.D.

● Infrared absorption measurement technique.   CE

CM20.9021   CM20.9061

FIGURE 4–22 Water and other contaminants monitoring systems
(Source: Courtesy of Parker Hannifin Corp. Hydraulic Filter Division, www.parker.com.)

many centuries that these heavens have revolved they would have been consumed by their own immense speed of every day . . ."

Dust particles in pneumatic conveyors and suspended particles in hydraulic systems can have a devastating effect of the performance and life of the equipment and can lead to increased energy consumption.

Tribology studies show that these particles, even soft materials such as wood dust, can cause erosive wear, especially when traveling around bends and under conditions in which materials are sliding against each other. The effect of these frictional forces and the extent of the damage is, to a great degree, a function of the physical properties of the two surfaces that are in relative motion to each other. Three general scenarios are possible: (1) a hard metal may be sliding against a soft surface, (2) a soft surface may be sliding against a hard one, or (3) metals with similar properties may be sliding against each other. Studies show that, even in the most favorable conditions, some surface damage will occur even with lightly loaded, well-lubricated surfaces.

Tribological failures are associated with bearings that can be involved in a complete shutdown of a fully integrated steel mill, resulting in hundreds of thousands of dollars in lost production and time. These economic losses necessitate the practice of failure prevention instead of failure correction maintenance.

4.2.4 Thermography

Various methods and tools for temperature sensing, such as pyrometers, thermocouples, and heat sensitive tapes and paints, are used in the predictive maintenance arena. Among these techniques, *infrared (IR) imaging* is one the most versatile and widely used methods for detecting surface temperature variances caused by abnormal or uncharacteristic conditions.

Infrared imaging (IR)   Heat from physical, chemical, or electrical abnormalities can be detected with infrared imaging and is usually a harbinger of a more serious condition requiring early and preemptive corrective actions. Mechanical wear, unbalanced and misaligned components, and disintegrated or inadequate lubricants all generate heat. Temperature increases usually can be detected using infrared imaging before vibration levels caused by these conditions can reach a detectable point.

Loose or corroded connectors or any other condition that can impede currents and increase electrical resistance, causing loss of energy and creating potential dangers and failures, can be detected by the use of thermography and infrared imaging. Loosely fastened or oxidized connectors and frayed electrical lines reduce the effective surface area for conductance and result in an increase in the surface temperature caused by increased resistance. Such temperature increases also occur in overloaded circuits, fuses, transformers, and so on, which can result in failures, fires, and other catastrophes.

Infrared imaging also can be used to locate leaking underground pipes, steam and gas leaks, and other sources of energy and material losses; and it can be used to reveal sources of energy loss from buildings and industrial facilities.

Infrared imaging and thermography is based on the fact that all objects at temperatures above absolute zero ($-273.15°C$ or $-459.67°F$) radiate infrared energy. The intensity of this radiation increases with the surface temperature of the object. The science of infrared thermography deals with detecting this radiation from a surface, and then processing it, converting it into an image, and interpreting the image to identify the "hot spots" and their possible sources.

Either direct temperature reading or visual analysis of the thermographs can identify hot spots or surfaces. Color thermographs usually are coded so that increases in the surface temperature progress from blue to green to yellow to orange to red, and finally to white, with blue representing the coolest and white the hottest areas of the surface. Each color represents a specific range of temperatures. On black-and-white thermographs, darker regions correspond to cooler areas, whereas hotter temperatures are represented by lighter colors and white.

Newer infrared imaging equipment resembles cameras in terms of weight and size (see Figure 4–23 and 4–24). They can be transported easily and are used by pointing at the area or surface to be analyzed. Obtaining an image is as simple as snapping a picture. Only minimal

FIGURE 4–23 Camera
(Source: Courtesy of Sierra Pacific Innovations, www.x20.org.)

FIGURE 4–24   Thermal imager
(Source: Courtesy of Sierra Pacific Innovations, www.x20.org.)

training is required to operate the equipment, but in order to interpret the images accurately, a higher level of knowledge and training is needed. A thermograph, or a heat image, from suspect equipment reveals moving mechanisms that are under stress due to friction or electrical components that exhibit overloaded conditions or abnormal resistance levels that eventually may become faulty. These images exhibit detectable temperature increases for faulty or stressed components that are above the normal and ambient temperature baselines. Thermography is a noncontact, nondestructive, and nonintrusive technique; therefore, predictive maintenance activities using infrared imaging do not interfere with plant operations.

The process can detect small changes in component temperatures caused by undesirable and potentially dangerous conditions. As with vibration analysis techniques, incremental changes in the component temperature indicate different stages of problem development and may require various preventive measures. Table 4–5 displays an example of how temperature deviations from the expected baseline in an electrical component might be interpreted and the possible recommendations to avert failure.

IR applications and success stories   The success stories associated with infrared imaging as a predictive maintenance tool are numerous. Peach Bottom Atomic Power Station in Philadelphia finds the technique

TABLE 4–5

Electrical Preventive Maintenance Reference Table

Problem classification*	Phase-to-phase temperature rise	Comments
Minor	1°–10°C	Repair in regular maintenance schedule; little probability of physical damage.
Intermediate	10°–30°C	Repair in the near future (2–4 weeks). weeks). Watch load and change accordingly. Inspect for physical damage. There is a probability of damage in the component, but not in the surrounding components.
Serious	30°–70°C	Repair in immediate future (1–2 days). Replace component and inspect the surrounding components for probable damage.
Critical	Above 70°C	Repair immediately (overtime). Replace component and inspect surrounding components. Repair while IR camera is still available to inspect after repairs are completed.

*With wind speed less than 15 mph and with load conditions greater than 50%.

Hint: Have an electrical contractor use a clamp-on ammeter to verify loading.

Source: Courtesy of Sierra Pacific Innovations, www.x20.org.

superior to bearing vibration analysis for detecting coupling conditions and degradation. Thermographic inspection is utilized as a reliable method for scheduling lubrication change and coupling replacements. Thermographic monitoring of pump couplings has virtually eliminated all catastrophic failures. After initial data gathering to establish a baseline, thermographic monitoring has accurately predicted the condition of the coupling and has helped avoid failures. Based on these analyses, lubrication and coupling replacements can be performed in a conveniently scheduled manner, reducing unplanned shutdowns and resulting in hundreds of thousands of dollars in savings.

Infrared thermography also is used to detect problem areas deep under the ground surface. The ability of the process to monitor large areas in a cost efficient, nonintrusive manner makes the technique an effective method for inspecting underground pipelines in chemical plants, water supply systems, steam and gas lines, and sewer systems. It is a convenient technique for detecting leaks and damage to underground conveyors of various substances. Since only the surface temperature of the object can be imaged, the thermograph from the ground surface may not necessarily indicate the temperature at the source. Based on thermodynamic principles and the thermal properties of the material, however, surface temperatures and infrared images can be used to calculate the subsurface temperatures and discover underground hot spots.

Refer to Figures 4–25 through 4–34 to discover and appreciate some interesting uses and applications of infrared imaging.

Other applications of infrared thermography include its use in the construction industry, and another interesting example is the use of infrared imaging to detect, the moisture in ancient historical sites, architectural works of art, and other masterpieces, in order to retard further deterioration of these structures. Infrared imaging provides an effective and economical method for monitoring thermal and hydrological parameters of masonry structures and artifacts without inflicting physical or esthetic damage to the structure since the process is nonintrusive.

Lest we neglect the sport enthusiasts, thermography also has been used by engineers and designers of ski equipment. Coefficients of friction between snow and ski, deformation characteristics of snow under pressure, and other interactions between snow and a skier's equipment all interest the equipment designers. Infrared thermography is one technique used to study these forces. As the skis glide over snow, the temperature increases as the result of the friction between the snow and the

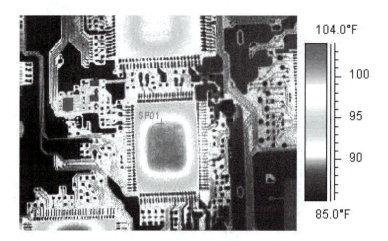

FIGURE 4–25 PC board
(Source: Courtesy of Sierra Pacific Innovations, www.x20.org.)

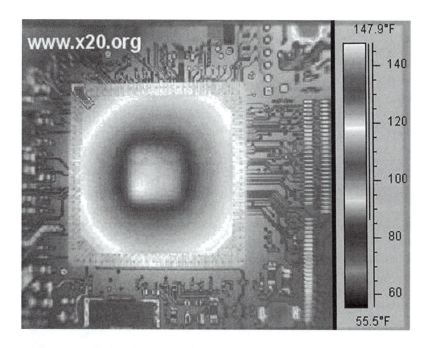

FIGURE 4–26   PC board
(Source: Courtesy of Sierra Pacific Innovations, www.x20.org.)

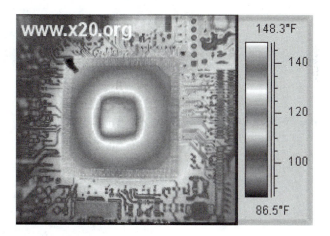

FIGURE 4–27   PC board
(Source: Courtesy of Sierra Pacific Innovations, www.x20.org.)

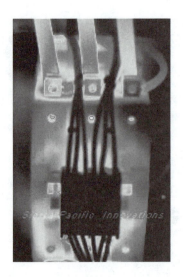

FIGURE 4–28   Electrical contacts
(Source: Courtesy of Sierra Pacific Innovations, www.x20.org.)

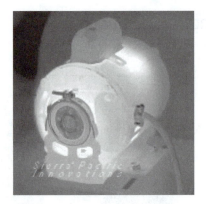

FIGURE 4–29   Motor
(Source: Courtesy of Sierra Pacific Innovations, www.x20.org.)

FIGURE 4–30 Storage tanks
(Source: Courtesy of Sierra Pacific Innovations, www.x20.org.)

FIGURE 4–31 Power plant steam tunnel
(Source: Courtesy of Sierra Pacific Innovations, www.x20.org.)

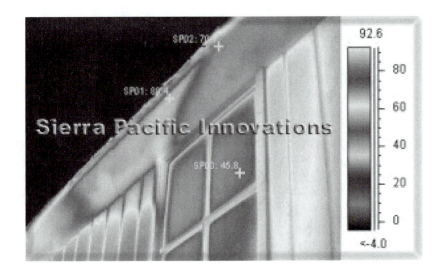

FIGURE 4–32   Energy audit
(Source: Courtesy of Sierra Pacific Innovations, www.x20.org.)

FIGURE 4–33   Energy audit
(Source: Courtesy of Sierra Pacific Innovations, www.x20.org.)

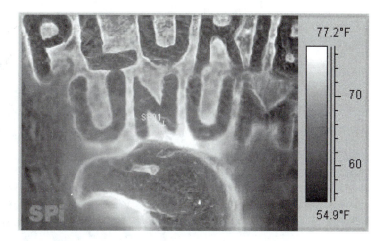

FIGURE 4–34   Do you recognize this? (U.S. quarter)
(Source: Courtesy of Sierra Pacific Innovations, www.x20.org.)

skis. Thermal analyses of the heat dissipation give clues for the ski base designs, shapes, and wax specifications.

IR equipment   Small, handheld devices, such as the one shown in Figure 4–35 using laser beam technology, also are available for temperature monitoring. Similar in size and shape to handheld barcode scanners used in retail stores, these devices can be used to measure the surface temperature of a moving or an inaccessible object accurately and quickly.

Less expensive methods of temperature detection and monitoring are available. For example, adhesive tapes, paints, and tags can be placed on equipment for this purpose. These objects have the ability to numerically register and display equipment surface temperatures, or through the use of strategically placed probes, exhibit the internal temperatures. Special tags, paints, or alarms also have the ability to display a range of color changes to visually enhance the detection of temperature variation and can sound alarms when the temperature reaches a critically low or high value. One advantage of these techniques is that they provide continuous monitoring of the system rather than the intermittent inspection afforded by thermographic procedures such as infrared cameras. Irreversible monitoring labels and tapes have the ability to maintain a record of the highest level of the surface temperate for future reference. Figure 4–36 displays a series of irreversible temperature recording labels; Figure 4–37 shows a 7-day, 24-hour portable

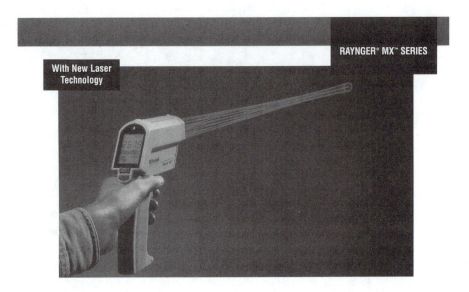

FIGURE 4–35   A handheld laser beam temperature monitoring
device
(Source: Courtesy of Raytek Corporation.)

time-temperature recorder; and Figure 4–38 illustrates a handheld precision probe thermometer.

4.2.5   Ultrasound Techniques

Ultrasonic frequencies are shortwave directional signals that are beyond the normal hearing range but can be detected by various instruments, and their intensity can be measured. All working equipment
produces characteristic ultrasound frequencies or "sonic signatures."
Changes in these sonic signatures signal changes in the working condition of the equipment and predict potential failures. Ultrasound frequencies also are emitted by leaks from hydraulic and pneumatic pipes,
steam traps, valves, and heat exchangers, as well as from electrical arcing and coronas caused by worn and frayed conductors, shorts, and
other dangerous abnormalities. Ultrasound frequencies often can be
detected before vibration or temperature levels from adverse conditions reach detectable stages. For example, NASA researchers have
found that potential bearing failures can be detected by ultrasonic
means long before vibration or heat detection techniques can be effectively used.

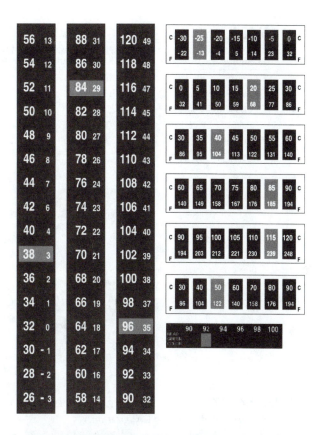

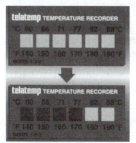

STANDARD 110 SERIES has six indicators which cover a range of 50°F on each model.

Model 110-10-13 is a 5°F increment label which covers a range of 100°F to 130°F (38°C to 54°C) with six temperature sensitive indicators.

FIGURE 4–36 Irreversible temperature recording labels
(Source: Courtesy of Telatemp Corporation.)

MODEL AND RANGE SELECTION TABLE

TEMPERATURE RANGE IN °C	24-HOUR SPAN MODEL NO.	7-DAY SPAN MODEL NO.	TEMPERATURE RANGE IN °F	24-HOUR SPAN MODEL NO.	7-DAY SPAN MODEL NO.
-40°C to +70°C	NA	T615.47CB	-40°F to +160°F	T615.41FB	T615.47FB
-35°C to +20°C	T615.41CC	T615.47CC	-30°F to +70°F	T615.41FC	T615.47FC
-15°C to +40°C	T615.41CE	T615.47CE	0°F to +100°F	T615.41FE	T615.47FE
-5°C to +50°C	T615.41CG	T615.47CG	+20°F to +120°F	T615.41FG	T615.47FG
-5°C to +105°C	T615.41CA	T615.47CA	+20°F to +220°F	T615.41FA	T615.47FA

FIGURE 4–37   A portable time-temperature recorder
(Source: Courtesy of Telatemp Corporation.)

Ultrasound categories   Ultrasound technology includes two major categories. One technique involves using a transducer that emits high-frequency ultrasonic waves directed toward an object or used to flood a shell or a part cavity. Echoes are collected, which are analyzed to reveal important information about the object. This technique can reveal changes in material properties such as thickness (wear), pits, cracks, voids, corrosion, and a host of other problems. This technique also can be used to reveal leaks in pipes or other containers. Since ultrasound frequencies lack the energy to penetrate solid walls, they can only "leak" through the holes. Scanning the surface will locate these holes.

The second technique involves the detection of the ultrasound frequencies generated by a source. Unlike audible sounds, which can vibrate solid surfaces or pass through them, making it nearly impossible to locate their sources, ultrasounds lack these qualities and are considered directional, which allows their sources to be located easily.

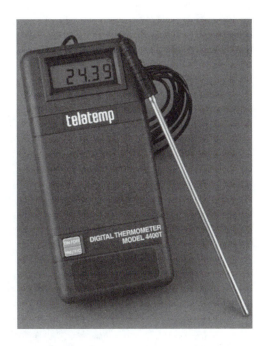

FIGURE 4–38   A handheld precision probe thermometer
(Source: Courtesy of Telatemp Corporation.)

Figure 4–39 shows a handheld ultrasound detector. The graph in Figure 4–40 illustrates the positive correlation between pressure and the ultrasound response, and the nonlinear relationship between the leak rate and the decibel reading is shown in Figure 4–41.

Ultrasound frequencies generated by damaged or worn pumps, gears, gearboxes, and bearings can be detected before vibrations reach detectable levels and cause any further damage to the equipment. Leaking gases as well as electrical arcs or coronas also emit ultrasound frequencies, as shown in Figures 4–42 through 4–44. As illustrated, these gaseous or electrical discharges can be detected quickly and economically by ultrasonic devices and the problems can be eliminated without failures and unplanned shutdowns.

Figure 4–45 displays another handheld condition monitoring device that utilizes ultrasound technology. The device, representing significant advances in the arena of preventive and predictive maintenance management and quality control, has a variety of uses in business and in the manufacturing and transportation industries. It is capable of detecting leaks and the rate of flow, mechanical wear, temperature measurement, rotational speed, noise levels, light intensity, and pH measurements. It simplifies the monitoring of bearings, valves, steam traps, and electrical arcing.

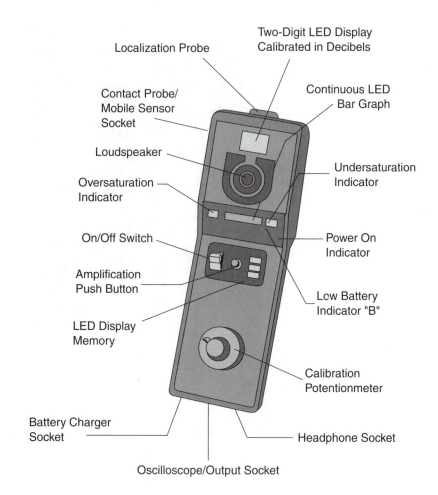

Localization Probe

Two-Digit LED Display
Calibrated in Decibels

Contact Probe/
Mobile Sensor
Socket

Continuous LED
Bar Graph

Loudspeaker

Undersaturation
Indicator

Oversaturation
Indicator

On/Off Switch

Power On
Indicator

Amplification
Push Button

Low Battery
Indicator "B"

LED Display
Memory

Calibration
Potentionmeter

Battery Charger
Socket

Headphone Socket

Oscilloscope/Output Socket

FIGURE 4–39   Ultrasonic detectors are like electronic stethoscopes,
hearing the sounds generated by a variety of physi-
cal, mechanical, and electrical phenomena

(Source: Reprinted with permission from the February 1998 issue of Plant Engineering
magazine. Copyright © 1998 by Cahners Business Information.)

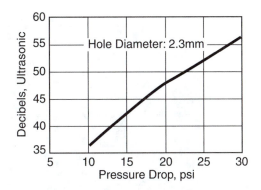

FIGURE 4–40   Ultrasonic versus pressure drop graph shows how
ultrasonic response increases with pressure

(Source: Reprinted with permission from the February 1998 issue of Plant Engineering
magazine. Copyright © 1998 by Cahners Business Information.)

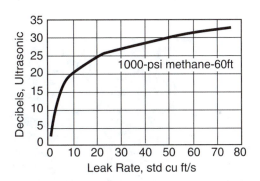

FIGURE 4–41   Leak rate versus ultrasonic graph shows the nonlinear relationship between decibel readings and leak
rate

(Source: Reprinted with permission from the February 1998 issue of Plant Engineering
magazine. Copyright © 1998 by Cahners Business Information.)

FIGURE 4–42   Ultrasonic instrumentation pinpoints compressed
air leaks fast, before energy is wasted and quality
affected adversely
(Source: Courtesy of Alan S. Bandes, UE Systems, Inc., Elmsford, NY.)

FIGURE 4–43   A user aims the ultrasonic device along the piping
while listening through headphones for the sound
of gas leaks
(Source: Courtesy of Alan S. Bandes, UE Systems, Inc., Elmsford, NY.)

FIGURE 4–44   An experienced user learns to adjust the sensitivity
dial on the instrument and once a reading is obtained,
the operator can consult a comparison flow-rate
chart to determine an approximate leak rate

(Source: Courtesy of Alan S. Bandes, UE Systems, Inc., Elmsford, NY.)

FIGURE 4–45   A versatile and portable condition monitoring
system
(Source: Courtesy of SDT North America, Inc.)

QUESTIONS

1. Define predictive maintenance and contrast it with other types of maintenance.

2. For an organization to be considered world class, what level of its maintenance activities should be of a predictive nature?

3. Predictive maintenance is said to be quantitative in nature. What does that mean, and how does that differ from PM?

4. What are some PDM methods?

5. How does vibration affect the useful life of a product?

6. What are some of the factors that determine the acceptable levels of vibration for a given piece of equipment?

7. Use Table 4–2 to answer the following questions:
   a. What are the velocity range limits (in mm/s) for a small machine in the "A" severity level?
   b. What are the velocity range limits for a medium machine in the "C" severity level?
   c. What are the velocity range limits for a class IV machine in the "B" severity level?

8. Define tribology and explain its significance in PDM.

9. How can oil analysis help in PDM?

10. What does the discovery of silicon in oil indicate?

11. What is the purpose of thermography and what role does it play in PDM?

12. Explain some of the applications of ultrasound in predictive maintenance.

13. What is pH and on what scale is it measured?

14. Can the pH of a solid be measured? How?

15. What is infrared and what role can it play in PDM?

16. What is the purpose of irreversible temperature labels?

5

NONDESTRUCTIVE TESTING AND EVALUATION

Overview

Objectives

At the completion of the chapter, students should be able to

- Understand various nondestructive evaluation techniques.
- Understand various basic applications of NDE to PDM techniques.

5.1   INTRODUCTION

Any testing and evaluation techniques that do not result in destruction of the specimen or product being tested can be referred to as nondestructive testing methods. Such techniques are particularly popular when the destruction of the test sample is not a desirable option, such as when testing expensive parts on production equipment.

Nondestructive testing (NDT) and nondestructive evaluation (NDE) techniques have been around for decades and have been used extensively in the field of quality control and quality assurance. Any method of testing and evaluating material properties in order to detect defects in engineering structures without impairing future use or altering material integrity can be classified as a nondestructive test.

The field of nondestructive testing has no clearly defined boundaries. Tests can range from a simple visual inspection of part and equipment surfaces to more sophisticated methods such as radiology and magnetic particle inspection, among others. NDE is concerned with a wide range of practical industrial and manufacturing problems, from simple trouble detection to more complex prediction of equipment and systems behavior. NDE methods are important in detecting flaws that can adversely affect the safe and reliable operation of equipment and the operational effectiveness of the plant.

Most common among the nondestructive testing techniques are ultrasonic methods, thermography, radiography, Eddy current tests, liquid penetrant tests, magnetic particle tests, and holography. Ultrasonic and thermography techniques were presented in Chapter 4. Discussions of each of other methods follows.

5.2   EDDY CURRENT TESTING

This nondestructive testing and evaluation technique provides a method to quickly and effectively test critical parts that might be operating in a hazardous or harsh environment without requiring direct contact with the test object. Eddy currents are induced directional flow of electrons in the test object under the influence of an electromagnetic field. An alternating current is passed through an exciting probe. The test object is placed near the probe, resulting in the flow of eddy currents in the test object, given that the test object is an electrical conductor, as shown in Figure 5–1. A pickup coil is used to measure the flow in the test object and any changes due to magnetic impedance. Any measured impedance in the flow of eddy currents due to bending of the flow lines or leakage of the electromagnetic flux points to the presence of flaws or other nonconformities in the test object. Figures 5–1a and 5–1b compare the resistance in two similar test objects. In Figure 5–1b, the jump in the internal resistance is due to the impedance caused by a surface crack in the part.

The test is based on comparison of the eddy currents in the test object with predetermined standards. The behavior of the eddy currents in test objects is dependent on the magnetic properties of the object. For example, aluminum is a nonmagnetic element, whereas iron compounds such as ferric steel are magnetic. It is therefore important to understand the effect of material properties on the eddy currents.

Eddy current testing offers a 100% accurate, high-speed, and reliable inspection of round and cylindrical objects such as bearings, tubes, and pipes. In earlier stages of development, eddy current testing was limited to round and cylindrical objects because special part geometry

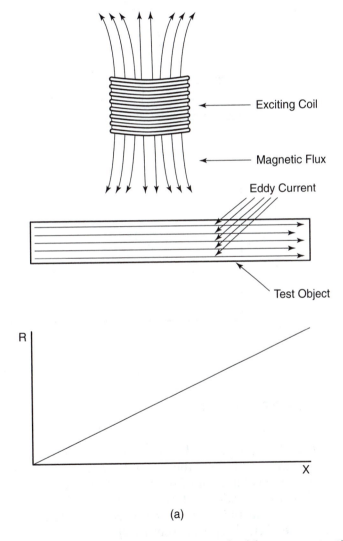

Exciting Coil

Magnetic Flux

Eddy Current

Test Object

(a)

FIGURE 5–1　A schematic presentation of eddy current testing
(continued on next page)

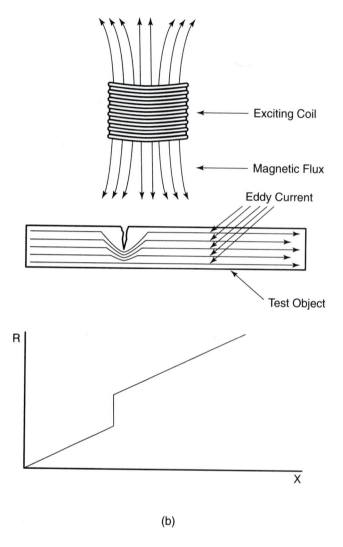

(b)

FIGURE 5–1   *(continued)*

(irregular shapes and configurations) interfered with the flow of eddy currents and limited the use of this technique. But recent research has led to the development of highly sophisticated exciting and pickup probes that have broadened the application of this NDT/NDE method.

The technique is widely used to locate and determine the size and shape of cracks, pits, and other defects in test objects. Current research is attempting to apply eddy current testing methods to help understand the behavior of multiple cracks on test objects under severe operating conditions such as chemical interactions, high temperatures, and oxidations. This has enabled researchers to predict the service life of structures containing multiple cracks and avert catastrophic failures.

One of the more significant uses of eddy current testing is for in-service inspection of tubes in steam generators and heat exchangers in conventional and nuclear power plants. The test offers high-speed inspection with a high degree of detection, and since direct contact with the test object is not required, it can be performed automatically.

Continuous analytical and experimental research involving the shape and size of the probes and induction methods has significantly improved flaw detectability. The flaw detection process not only has been significantly improved, but notable progress also has been achieved in precise evaluation of the flaw, including determination of flaw size, orientation, characteristics, and origin. These enhancements continue to diversify the use of this method of nondestructive testing in the preventive and the predictive maintenance arena.

THE INSPECTION OF PRESSURIZED WATER REACTORS

According to a 1999 report by the International Energy Agency, 103 nuclear power plants are currently operating in the United States, producing nearly 20% of all the electricity generated in this country. These statistics indicate that nuclear energy is one of the main sources of electricity in this country, second only to coal and ahead of natural gas, with the former producing approximately 52% and the latter less than 10% of all the electricity generated in the United States.

Whether nuclear power is the answer to our ever-increasing and insatiable thirst for energy is a topic that is certain to cause heated debate and contentious disagreement. But what is certain and should enjoy unanimous agreement is the fact that nuclear power plants do require the utmost vigilance and proper maintenance to ensure their safe operation. Catastrophes such as Three Mile Island (1979, U.S.A.), Chernobyl (1986, former U.S.S.R.), and Tokaimura (1999, Japan) are all frightening and familiar stories.

One important application of eddy current testing is for the inspection of steam generator tubing in nuclear power plants. These generators act as heat exchange units that transfer thermal energy from the nuclear reactor to the turbine. The extremely harsh environment causes the tubes of steam generators that carry the radioactive coolant to suffer from degradation that causes cracks and pits. It is critical that surface cracks in tubes of pressurized water reactor (PWR)–type nuclear

(continued on next page)

plants are detected before they become too large.

Currently, there are over 20 pressurized water reactor plants operating in Japan. Each plant has approximately three to four steam generators, with nearly 3,000 tubes per generator. The entire length of each tube (roughly, more than one-quarter million tubes!) needs to be inspected annually to assure the structural integrity of the tubes and prevent leakage of radioactive steam and coolant into the environment and other related catastrophes.

Eddy current testing, utilizing a special type of noncontact probes, provides for a fast scanning speed with a high degree of detectability. Continuing research has resolved the problem of durability of the probes to a large degree and has reduced the lack of uniform sensitivity.

Source

Kurokawa, M., R. Miyauchi, K. Enami, and M. Matsumoto. "New Eddy Current Probe for NDE of Steam Generator Tubes," *in* D. Lesselier and A. Razek, eds., *Electromagnetic Nondestructive Evaluation*, vol. III, p. 57. Washington, DC: IOS Press, 1999.

5.3 RADIOGRAPHY

Techniques that use penetrating radiation to obtain an internal latent or shadow image of a solid object are generally referred to as radiography. The most common types of radiation used for this purpose are X rays and γ rays (gamma rays). Although the resulting images are usually recorded on film similar to the film used in common photography, in real-time radiography, or radioscopy, the images are displayed on television monitors for real-time inspection and viewing.

Similar to ordinary light, X rays and γ rays have the ability to expose photographic film and produce latent images. Prolonged and more intense exposures to these radiations create a darker image, as with ordinary light. This ability, coupled with the ability of X rays and γ rays to penetrate solid objects, make it possible to produce images of the internal structure of various objects.

For the most part, X rays and γ rays have the same characteristics. Their extremely short wavelengths enable them to penetrate substances that reflect or absorb ordinary light. The major distinction between X rays and γ rays is based on their source rather than their behavior. X rays are emitted when an electron beam strikes a solid object, usually a tungsten target in a vacuum tube. Since a very small amount of the energy (approximately 1%) is usually converted into X rays, these systems are considered inefficient generators. Natural sources of γ rays are disintegrating radioactive substances such as radium. Cobalt-60, another source of γ rays, is an artificial source.

The contrast or optical intensity of the radiograph depends on the amount of radiation that has been absorbed or has been allowed to pass through the object and reach the film. As the radiation passes through the test object, in areas where the material density or thickness *absorbs* a larger amount of radiation, less radiation is allowed to pass through, so less radiation reaches the film, causing a lighter shadow. In areas where voids, cracks, or other factors have diminished the relative density of the object, however, more radiation is allowed to pass through and reach the film, subsequently creating a darker image. Pronounced variation in the optical density caused by various degrees of absorption of the radiation can point to a lack of homogeneity of the chemical composition, nonuniform density, internal cracks, voids, discontinuities, and other flaws. Given that X rays and γ rays can pass through different substances at different speeds and with varying degrees of absorption, radiography can be used for the nondestructive testing and evaluation of forgings, welds, castings, and many other metallic and nonmetallic fabricated parts. The inspection process requires comparison of the results with predetermined standards to determine abnormalities in the test object.

State-of-the-art X-ray technology is available in many sizes. Portable equipment provides small, tabletop cabinets with superior real-time image quality. Figure 5–2 displays some advanced systems. Application of X-ray technology is also quite diversified and can be found in most areas of business and industry as well as science and technology. Figure 5–3 illustrates a few common examples of application of this technology.

Figure 5–3a probably represents the most widely known use of X-ray technology—discovery of hidden objects. Contraband, drugs, weapons, and other concealed objects can be detected easily with X rays.

Figure 5–3b shows the inside of a weapon. Often a weapon may be rusty or corroded and a radiograph can provide valuable information regarding the firing condition of the gun. Radiographs have been used in court cases to illustrate differences in the inner surfaces and the firing conditions of two seemingly identical weapons.

Ballistics investigation is another arena in which radiography plays an important role. The area surrounding a bullet hole is shown in Figure 5–3c. X-ray technology can provide useful information about the type of bullet, firing distance, and direction of the projectile. An X ray can separate the background material from the lead and powder particles so that only the splatter pattern caused by the discharge of the weapon can be observed.

The bone structure of a mouse's leg is shown in Figure 5–3d. The use of X ray and radiography has long been established for the study of the internal organs and for detection of bone fractures and other anomalies.

Faxitron Model 43855F
with real-time imaging package

Radiograph of aluminum
casting with porosity

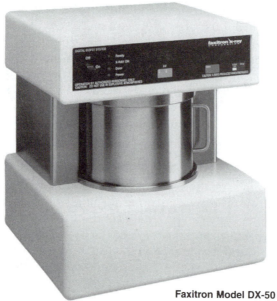

Faxitron Model DX-50

FIGURE 5–2   State-of-the art X-ray equipment
(Source: Courtesy of Faxitron X-Ray Corporation.)

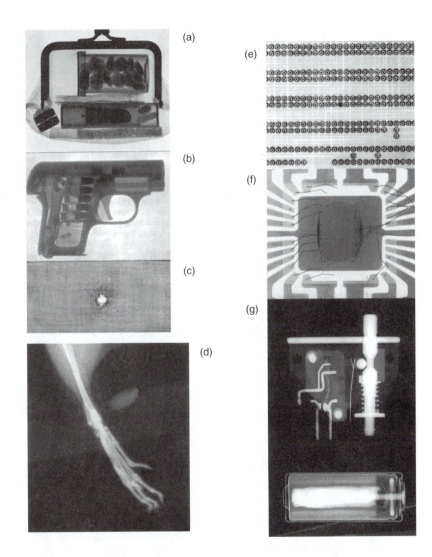

FIGURE 5–3 Some applications of X-ray imaging
(Source: Courtesy of Faxitron X-Ray Corporation.)

Figure 5–3e displays a multilayer PC board. X-ray inspection can determine if the internal layers are properly aligned and whether or not drilled holes are properly positioned in relation to inner layer connectors. Any defects, cracks, or breaks in metal plating also can be determined.

Transistors, integrated circuits, and hybrid circuits can be X-rayed to look for defects such as improper bonding to the chip and porosity or poor bonding to the substrate. A semiconductor is shown in Figure 5–3f.

X rays can be used to inspect canned and potted assemblies, switches, and relays for proper location of internal parts, lead connections to external contacts, and welds, as shown in Figure 5–3g.

5.3.1   Radioscopy

In certain instances, when the incoming X rays have sufficient energy, they may cause the object itself to generate X rays and become radiant. This is known as fluorescent radiation. Radioscopy, also known as fluoroscopy, allows the observation of the object in real-time on a screen rather than by recording the image on a film for delayed inspection. The technique can be used for rapid and economic inspection of moving objects. The object may be inspected in real-time from various angles. The cost of film and film processing and handling also is eliminated.

5.3.2   Neutron Radiography

Neutrons are one of the smallest particles that make up an atom. Neutron radiography complements X-ray techniques with its ability to produce images of objects and materials that are transparent to X rays. Absorption characteristics of neutrons vary significantly from those of X rays. Substances with high atomic numbers, such as lead (Pb, 82) heavily absorb X rays; conversely, hydrogen (H, 1), which is totally transparent to X rays, has a high absorption rate for neutrons. Therefore, neutron radiography provides a greater contrast and a better imaging technique for organic substances such as leather, plastics, and fluids.

5.3.3   Tomography

Tomography is a special area of radiography that provides information in three dimensions; therefore, the location of any flaws and defects can be determined. In conventional radiography, as the X rays pass through the length of the object, varying degrees of absorption occur along the path due to lack of uniformity caused by cracks, flaws, and so on, resulting in the contrasts that are shown on the film. However, no information about the relative distance of the defect along the path of the X rays is provided. Tomography reveals the position of the defect more precisely by identifying and isolating the plane in which the defect is located. The process is analogous to creating multilayer slices of an object, creating a three-dimensional image of the object without actually cutting it.

5.3.4   Magnetic Resonance Imaging

Magnetic resonance imaging, commonly referred to as MRI, is a tomographic imaging procedure that allows the production of multidimensional images of the internal structure of physical and chemical structures in real-time on a television screen without invasive methods. MRI technology has made significant contributions to science and medicine by aiding in the early detection and diagnosis of many types of illnesses and internal abnormalities. Some methods of fluoroscopy require high-energy ionization radiation in order to create an on-screen image, whereas other imaging techniques may require that the object be injected with radioactive isotopes in order to generate a signal. MRI is capable of generating multidimensional images in any orientation without the ionization requirements or the use of radioactive materials, hence avoiding both hazardous conditions.

5.3.5 Quantitative Radiography

Conventional radiography is a qualitative evaluation by nature. The contrasts in the images are indicative of lack of homogeneity, suggesting the presence of cracks, flaws, and other nonconformities. Continuous research in radiography not only has expanded the use of this methodology through improved equipment, film production, and development, it also has significantly improved the precision of this nondestructive testing technique.

One area of advancement is in quantitative radiology. Current manufacturing and industrial demand for advanced materials has led to the development of sophisticated synthetic and composite materials. These materials often contain such slight variations in density that they are often undetectable to qualitative radiography. Quantitative radiography has made it possible to digitize and display a quantitative image of an object to allow for a numerical comparison with set standards. Many parts with varying composition and characteristics can be digitized with quantitative radiography.

5.4 LIQUID PENETRANT TESTING

Liquid penetrant inspection techniques provide an effective, quick, and a reliable method for detecting surface cracks in metallic as well as nonmetallic objects. The application of this procedure is simple, requires very little training, and the technique can be used to detect extremely small open surface defects. Liquid penetrant testing has proven to be more reliable than radiography and is a more economical means of detecting surface openings and cracks than ultrasound techniques. Since the procedure is also applicable to nonferrous and nonmetallic materials, it is more versatile than magnetic particle and eddy current testing.

Blacksmiths and iron workers have employed this technique to detect surface cracks by dipping an object in oil, wiping the surface clean, and then covering the surface with fine chalk suspended in alcohol. Once the alcohol evaporates, the surface remains covered with a white coating of chalk. Residual oil deposited in fine surface cracks seeps out when the object is vibrated, producing stains in the white chalk coating and making the cracks visible. This method, which became known as the oil-and-whiting technique, also was used in railroad plants to detect surface cracks in locomotive parts and train wheels.

The procedure for modern liquid penetrant testing follows these basic steps:

1. The surface of the test object must be thoroughly cleaned and free of oil, dirt, or any material that would interfere with the permeating action of the penetrant.

2. The liquid penetrant, which contains a dye, is applied to the surface of the object by dipping, submerging, or spraying. A sufficient "dwell" time allows the penetrant to permeate into surface cracks and openings, as shown in Figure 5–4a.

3. The excess liquid is then carefully removed and the surface is wiped clean. The excess penetrant must be completely removed, but care must be taken so that the penetrant is not extracted from any cracks.

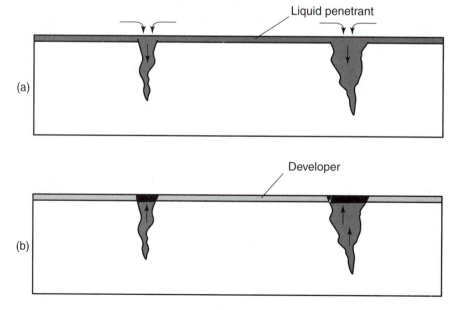

FIGURE 5–4   Illustration of the liquid penetrant action

4. The next step involves the application of a developer onto the surface. The developer may be a fine powder, or as in the example of oil-and-whiting, it may be suspended in a special solution. Given sufficient dwell time, the developer draws the penetrant to the surface by capillary action, making the surface cracks visible. Figure 5–4b shows the highlighted areas in the developer due to the presence of the penetrant in the surface cracks.

Modern dye and penetrant molecules applied in liquid penetrant testing are small enough to detect practically all surface cracks. The technique, however, is not appropriate for subsurface and internal cracks and defects.

5.5   MAGNETIC PARTICLE TESTING

The use of magnetic particle testing is based on the fact that steel or other ferric material will align itself on magnetized surfaces where magnetic flux lines enter or leave the object. Magnetic particle inspection is a nondestructive testing method for detection of surface or subsurface discontinuities and defects such as cracks in ferrous materials. A series of imaginary lines of force are formed around a magnetized steel bar. The pattern of these lines can be plotted by sprinkling fine iron filings on a clean magnetized steel bar. The iron particles become magnetized and will form a link along the magnetic lines of force as shown in Figure 5–5. A more clear visual display can be produced if the filings contrast in color with the magnetized surface.

A crack, inclusion, or any discontinuity on the surface of the object will disturb these lines of force in such a way that parts of the steel bar

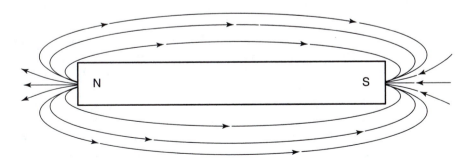

FIGURE 5–5   A pattern of magnetic force lines

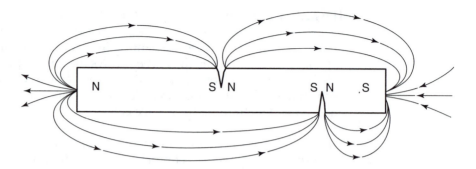

FIGURE 5–6   Magnetic force lines resulting from a cracked surface

on each side of the crack will behave as separate magnets, forming opposite magnetic poles on each side of the discontinuity, as shown in Figure 5–6. This distortion in the magnetic force lines is called a magnetic leakage field.

Magnetic particle testing therefore depends on the detection of a magnetic leakage field. The powdered material is attracted to and forms patterns where magnetic flux lines enter or leave the object; therefore, these patterns are formed at points, lines, or separations in the objects, making the detection of breaks, cracks, or other defects possible. The technique can be used to inspect for cracks or other defects caused by material fatigue in aircraft engines, crankshafts, and axles, to name a few examples. The use of magnetic particle testing was made mandatory at the Indianapolis Speedway in the mid-1930s.

5.5.1   Magnetic Particle Classification

Magnetic particle testing can be divided into dry and wet classifications. In the dry testing method, dry powder is sprinkled on the test object. The surface must be dry and free of rust, oil, and dirt. Particle size and shapes are selected with care. Particle types determine their magnetic behavior, whereas the size and shape of the particles affect their fluidity. Colored or fluorescent particles make it easier to visually inspect for magnetic leakage patterns.

In the wet testing method, the magnetic particles are suspended in oil or some water-based liquids, and the test piece is submerged in the liquid. The object must be free of dust, loose rust particles, scaling, and it preferably should be degreased prior to the test. Dry testing is more appropriate for detecting defects and cracks beneath the surface of objects. In the case of fine surface cracks, wet testing is the method of choice.

5.6 HOLOGRAPHY

Holography, from the Greek word *holos*, meaning whole or complete, is the science of creating and recording a three-dimensional image of an object. Conventional photography creates two-dimensional images from three-dimensional objects, but holography preserves the object's three-dimensionality. In the early stages, holography was used in the arts to create beautiful three-dimensional scenes and pictures. Today, special holographic techniques called interferometric holography or holographic interferometry have many applications in engineering and in material testing and evaluation.

Conventional holography utilizes a split laser beam of coherent light to create one three-dimensional image of the object. Interferometric holography is similar to conventional holography except that in this method, two holograms are created. These two three-dimensional images of the object tend to occupy the same location in space at the same time, interfering with each other and creating sets of contour lines as a result. These lines are called interference fringes.

The ability to create two images of the same object and the resulting interference fringes make it possible to detect any slight variation in the images that are due to dimensional differences or the surface displacement of the object at two different states of the object. So any internal defects, flaws, surface cracks, or discontinuities will cause the object to respond differently to various external stimuli, such as thermal or mechanical stress, creating displacement and deformation in the surface of the object between the two exposures and creating irregularities in the fringe interferences. It is therefore possible to detect and locate any cracks, breaks, or other structural flaws in the object.

Interferometric holography provides for nondestructive evaluation of parts by obtaining information over the entire surface of the object rather than just at a point or points on the surface. It is a particularly attractive method if the physical environment or part characteristics make it difficult to establish direct contact with the object. Furthermore, the examination can take place in real time as the part is subjected to external stress such as vibration, mechanical loading, or thermal stress. The presence of surface cracks will cause a disturbance in the interference fringes at the location of these structural maladies. This technique can be used to detect very small cracks and to track their growth over time by observing the displacement in the surface of the object.

A combination of ultrasonic vibration with interferometric holography, called acoustical holography, provides an excellent means for detecting and measuring disjoints, voids, and cracks. Since these flaws can be made to vibrate easily, the structural displacements are readily recognizable by interference fringes. Important advantages of using acoustical holography instead of conventional ultrasound techniques are the method's use of lower frequencies and its ability to evaluate larger surfaces.

QUESTIONS

1. Explain NDT and state some of the techniques that are used in PDM.

2. What is the NDT technique used in the predictive maintenance process in Japanese nuclear power plants discussed in the case study?

3. Under what general category of NDT does X-ray technology fall, and what are some of its industrial uses in PDM?

4. To what NDT technique does the "oil-and-whiting" technique refer? Explain how the method works.

5. How is magnetic particle testing used to detect surface cracks in objects?

6. Would you recommend the use of magnetic particle testing for the detection of surface cracks in an aluminum object? Explain your answer.

6

IMPLEMENTING TPM

Overview

Objectives

At the completion of the chapter, students should be able to

- Define total productive maintenance and the steps necessary to implement it successfully.
- Calculate overall equipment effectiveness (OEE) and assess its individual parts as well as the whole.
- Identify the basic TPM activities and processes.

6.1   TOTAL PRODUCTIVE MAINTENANCE

Total productive maintenance (TPM) is a philosophy; it is a way of thinking; it is a culture. TPM was developed in Japan, a society in which teamwork is highly revered and individualism is not venerated to the same degree as it is in America. And therein lies the challenge of TPM implementation.

6.1.1   Successful Implementation of TPM

In order to successfully implement TPM, it is necessary to establish that TPM is not merely a program for the maintenance department but rather a corporate-wide equipment and resource management strategy for overall quality and productivity improvement. TPM is a strategy deeply rooted in teamwork, education and training, communication, ownership, and empowerment. It is possible to understand TPM in the same way as someone can understand playing a sport, such as football, by reading about it and watching others play it. Actually *playing* the game, however, requires a great deal of patience, practice, and of course, teamwork.

Review Tables 1–1 through 1–3, this time from a different perspective. According to the data, over 55% of the maintenance activities in this country are corrective. Only 33% and 13% are preventive and predictive or condition based respectively. So more than 55% of the time, the maintenance organization is waiting for equipment to break down, which means that a loss in quality and productivity *will* occur before any action is taken. Perhaps this approach is based on our country's historical pride in being great "fire fighters" and our proven ability to develop crisis management teams. The modern global economy and worldwide competition, however, require a shift in our expertise from great crisis managers to crisis averters as a matter of survival.

To understand and implement TPM is to understand the culture of change. TPM is involved with how to *avoid* a crisis in the first place. Instead of focusing on how quickly we can *react* to restore equipment and reduce downtime, TPM changes the focus to how to be proactive by increasing equipment reliability and maximizing uptime, based on the principles of preventive and condition-based maintenance. The implication of TPM, total *productive* maintenance, is that every part of an organization must work toward increases in productivity. Productivity and quality gains can be realized when they are built on the pillars of TPM.

The cornerstone of TPM implementation and practice is total participation and involvement of all personnel, especially the operating personnel. TPM requires a shift in the ownership of and responsibility for the equipment and its upkeep to the operators. The operators must have an active role in the basic maintenance of the equipment, which represents a shift to autonomous and self-directed maintenance. To achieve this shift in ownership and responsibility, a great deal of preparation and planning is required—necessary in planning for any journey.

6.2 A CHANGE IN THE CORPORATE CULTURE

TPM requires a change in the corporate culture. It is not a fad, and it does not occur overnight. In order for the process to be successful, long-term and complete commitment and participation are required from the management—at all levels.

Often employees are uncomfortable with change, and most do not believe that change is going to take place. For the past 30 years, corporate America has faced a parade of three-letter hypes (MRP, JIT, TQM, and so on) that have appeared, generated some excitement and some anxieties, and then faded away, often without much accomplished. So why should this three-letter newcomer be any different?

The level of success that a company has experienced with such ventures in the past is a good indicator of future performance. If the past experience has been a positive one, in which a team effort with the commitment and participation of the management has brought these projects to some reasonable fruition, then one can expect warm enthusiasm when TPM is introduced. On the other hand, if past projects have faded away, perhaps from a lack of serious long-term commitment from the management, then employee skepticism should not surprise anyone.

The concept of change also tends to be unsettling to most people. It has been said that the only one who likes a change is a baby with a dirty diaper. So, people will ask, why is the change needed, and how is it really going to be any different, and most importantly, how is it going to affect "me"? These questions need to be addressed and answered so that everyone in an organization understands how TPM will change the workplace in a positive way.

6.2.1 Deming's Legacy

The late W. Edward Deming, the father of modern statistical quality control and the founder of the Japanese quality revolution, introduced a 14-point system for creating and improving the quality of corporate culture. In recognition of his great contributions to Japan and the development of that country's quality systems after World War II, the Union of Japanese Scientists and Engineers instituted the highly coveted Deming Prize, which is awarded to individuals in companies who achieve the highest level of quality management practices. Each and every one of these 14 points is a valuable lesson, but a few of them are most applicable to this discussion.

According to Deming, management must create a statement of purpose and share it with all employees. The statement should clearly explain the changes that will be implemented, including what the objectives of the new practices are, why these changes are necessary, and how they will affect various individuals. He advocated eliminating fear, creating trust, and developing an environment that fosters innovation and empowerment, accomplished by clear, open, and honest communication. Deming strongly believed in removing communication barriers that prevent individuals from experiencing the pride of workmanship and self-improvement. He strongly advocated training and education. TPM is deeply rooted in these philosophies: Communicate the philosophy and purpose of TPM through open and clear communication and drive out the fear of all concerned individuals.

6.3 UNIONS AND TPM

TMP requires some maintenance and upkeep activities to move from the traditional maintenance department to the operator. A common and logical reaction to this kind of change is that it means more work for the operator and probably a cutback in the maintenance department. In a traditional union shop, this may appear to be a violation of job descriptions and classifications, which the union would oppose. The union also would be concerned about any reductions in number of employees. But if the purpose of TPM is to reduce the number of maintenance department employees, the strategy is doomed to failure from the start, whether the plant is a union shop or not.

The true purpose of TPM is to improve quality and productivity by increasing uptime and achieving zero defects with the aid and direct involvement of the operator, an operator who is well trained and qualified to take ownership of the equipment and actively participate in its maintenance and upkeep. Once union leaders and members understand TPM, they will not fear it and therefore they will not oppose it.

TMP implementation can and has been successful in union plants such as AT&T, DuPont, Ford Motor Company, Motorola, and Texas Intrument. The management and the union both must understand that the survival of the corporation, and therefore the job security of union members, depends on how successfully the company can compete in the global economy. Both the management and the union must realize that their adversarial relations of the past must give way to a spirit of mutual respect and cooperation for the benefit of the company and all its employees. The purpose of TPM is not to shift maintenance functions from the maintenance department to the operators to reduce the number of maintenance employees. Instead, TPM allows the operator to take an active role in equipment upkeep so that problems can be detected early and averted, which allows the maintenance department to concentrate on more proactive and predictive maintenance functions.

TPM can improve the overall quality of work life by improving the safety of the equipment and the workplace, a primary concern of any union, by creating a cleaner environment, and by upgrading the working conditions.

Union leaders and the membership should be involved in the early stages of planning for TPM. The adage that "If they are not with you, they are against you" is applicable. Therefore, from the earliest planning stages, the employees, union and nonunion workers, must be involved in the process. Institute a well-planned TPM education and training program. As the training phase of TPM progresses, utilize the trained employees as trainers for the subsequent classes. Once the union leaders and the membership see the advantages of TPM and how it is instrumental in improving employee skills and job security, once they see how TPM will improve the safety and the quality of their work life, and once they see that they are indeed partners in the process, they will be advocates instead of detractors.

6.4 THE TPM JOURNEY

As is the case with any passage, the TPM journey requires a clear plan and a road map. How do we know where we are going or that we have arrived if we do not know where we are coming from?

6.4.1 Introducing and Promoting TPM

Once corporate management has made a total and complete commitment to TPM, it is time to involve union leaders to introduce TPM and its goals and objectives to the employees. This can be done through seminars, meetings, and presentations, with or without help from professional consultants and presenters, depending on the availability of in-house experts and presenters. Be sure to get employees involved from the early stages of the process.

Employees who have been trained in the basic concepts of TPM can be utilized to form small committees that will communicate and introduce other employees to TPM in small and informal settings, to propagate the concepts and philosophies of total productive maintenance. These committees can clearly communicate management's commitment and participation in the process and address issues of concern.

6.4.2   Communicating the Goals and Objectives of TPM

Improving uptime, quality, and productivity through equipment maintenance and performance improvement are among the goals of TPM. TPM aims to eliminate unplanned equipment downtime, achieve zero defects due to equipment malfunction, and eradicate loss of equipment speed. These goals are achieved through operator-based, self-directed, autonomous maintenance activities. Basic routines such as cleaning, inspection, and lubrication significantly reduce and eliminate unplanned shutdowns with a minimal investment of time.

Well-kept and well-maintained equipment is the key to eliminating defects caused by equipment malfunction. Any deviation in equipment cycle time compared to the theoretical time is considered a loss in equipment speed and results in a loss of productivity. If the slower equipment is in line with other pieces of equipment, reduction in speed in one piece of equipment will result in loss of speed in the remaining equipment in line. A trained operator can detect these losses easily.

6.4.3   Assessing Current Conditions

It is necessary to establish the current equipment status and take an inventory of current skill levels of operators to plan successfully for the proper training of the personnel.

Assessing the current status of the equipment can be accomplished by several means. A review of existing and past work orders gives a good indication of the equipment status. Establish the percent of emergency work orders. Check equipment reliability, frequency and rate of failures, percent of downtime, the size of the backlog, and the allotted budget for the maintenance department; these are also good indicators of current condition.

Determining the condition of current equipment and the current effectiveness of the maintenance department may be surprising—and not necessarily in a pleasant way. But establishing the current status and effectiveness of the maintenance department and the equipment will highlight the need to change the status quo.

Evaluate each piece of equipment and rank it for overall condition and reliability using Table 6–1. Based on the team members' knowledge and the historical data, match the overall condition of the equipment with the various conditions given in the table and determine the rating for each piece of equipment. The table also lists a possible course of action for equipment in each rating category.

Once you have established the current condition of each piece of equipment, determine the percentage of the equipment that falls in each of the rating categories. How does the majority of your equipment rank?

TABLE 6–1

Equipment Condition Analysis

Rating scale	Condition	Possible actions
1 Poor	• Below all standards • Very difficult to operate • Unreliable • Very low OEE • Does not hold tolerance • No improvements done • Unsafe to operate • Very high scrap rate • No PM	**Requires immediate attention** • Scrap • Rebuild • Start PM • Improve function and safety • Clean up • Repaint • Hide
2 Fair	• Barely acceptable • Below standards • Not easy to operate • Limited capability • Dirty • Low OEE • High scrap rate • Very little PM	**Requires early action** • Rebuild • Improve function and safety • Improve PM • Clean up • Improve inspection
3 Average	• Meets requirements • Fairly reliable • PM done • But not in good condition • Some limited capability • Decent appearance • Average OEE • Average scrap rate	**Requires action** • Improve necessary functions • Improve inspections • Improve PM • Clean up • Don't let deteriorate
4 Good	• Reliable machine • Nice appearance • Very little scrap • All PM done • Some improvements done • Good OEE • Meets all standards	**Possible actions** • Fine tune PM • Keep inspecting equipment • Keep cleaning/lubricating • Improve where possible • Don't let deteriorate
5 Excellent	• Perfect condition • Looks brand new • Excellent capabilities • No scrap • Equipment improved • No breakdowns • Perfect PM done • Excellent OEE	**Use as example** • Show off to customers • Don't let deteriorate • Maintain perfect PM record • Keep perfectly clean

Source: Courtesy of Edward H. Hartmann, International TPM Institute, Inc.

Below average? Excellent? Poor? The answers to these questions should provide a clear picture of the current status of quality and productivity and to some extent, the effectiveness of the organization's maintenance program.

6.4.4   Assessing Current Maintenance Effectiveness

The effectiveness of current maintenance programs also can be measured, using the criteria in Table 6–2. For each area, utilization, methods, and performance, match your current operating status with the table entries and find the closest match. How would your organization's daily activities best match the entries in the utilization column? If the company faces frequent inventory stockouts that result in delays, or historical data is not available or not used for scheduling purposes, or there is no effort to coordinate the use of different skills of employees, then the company would earn a utilization rating of 40% at best.

Similarly, rate the organization's operating methods. A high degree of organizational planning, such as frequent use of the correct tools and equipment or highly qualified and well-trained personnel will result in a high rating in the methods category. But a lack of work instructions or standards, old and obsolete equipment, and poor planning will result in a low rating in this category. Using the same process, establish the company's rating in the performance arena.

The current level of the organization's maintenance productivity can be calculated by multiplying the three ratings:

$$\text{Productivity} = \text{Utilization} \times \text{Methods} \times \text{Performance}$$

Assuming that a given company is rated at 65%, 50%, and 75% in utilization, methods, and performance respectively, then

$$\text{Productivity} = 0.65 \times 0.50 \times 0.75 = 0.24 \text{ or } 24\%$$

It is a great incentive to set a realistic productivity goal and then take advantage of opportunities to attain it. It is also important to establish your current baseline prior to embarking on the TPM process so that you can measure your productivity gains. Success stories are the best morale boosters and help pave the way for future success.

Comparing the level of various maintenance activities (corrective, preventive, and predictive) of a particular organization with that of a world-class organization as indicated in Tables 1–1 through 1–3 or with a set of realistic goals and objectives established for the organization also provides an indicator of the amount and the extent of work that lies ahead.

TABLE 6–2

Current Maintenance Productivity

	Utilization	Methods	Performance
40% 50%	• Noticeable job delays and idle time • Informal material and parts control • Frequent stockouts • No record of used time • No skill coordination • No work planning • Historical data not used for scheduling • Many old or outdated drawings • No scheduling	• No job instructions • No industrial engineering effort apparent • No standards • Frequent "rework" required • Old equipment • Methods and planning work left to workers • Work considered too difficult to predefine • Very few "one-man jobs"	• Frequent revisions of job • Many interrupted jobs • Low workload level • Occasional good effort noticeable • No reporting system used • No supervisory training • Low skill level • Poor attitude
60% 70%	• Frequent delays and "walking around" • Some effort to control material and parts • Some jobs preplanned by foremen with minor follow-up • Unclear organizational lines and jurisdiction • Foremen estimate time usage • Informal skills coordination • Causes for delays unknown	• Only occasional work planning • Frequent group discussions on how to do job • Mild management interest in methods • Instructions only provided for large jobs • Some standard practices • Work methods evolved rather than planned	• Informal supervisory training • Job assignments unclear • Reasonably steady work effort • Vague job instructions • Future work uncertain • Danger of layoff • Time controls loosely followed • Supervisors rarely visit job site • Informal reporting system
80%	• Few delays or stockouts • Material requirements are preplanned • Formal planning and scheduling procedures used • Good management information available • Few complaints from requestors • Good control of costs and work backlog • Reporting system shows utilization	• Majority of jobs preplanned • Good planning of work • Personnel are methods conscious • Permanent industrial engineering group assigned • Frequent methods suggestions • Few changes in planned work • Correct tools and equipment often used	• Steady work effort • Businesslike operation • Pride in workmanship and knowledge • Clear job instructions • Good time controls • Supervisor knows status of all jobs • Reporting system shows performance and productivity

(continued on next page)

TABLE 6–2

(continued)

	Utilization	Methods	Performance
90% 100%	• Practically no delays • Required materials always available • All plannable jobs are preplanned • Management controls pinpoint problem areas • Sophisticated planning and dispatching of work	• Tools and equipment tops • Standard tools and procedures available • Some volume of repetitive work • High skill level of workers • All methods are preplanned	• Sound, working incentive plan • Well-trained supervisors • No recent labor trouble or strikes • Good backlog of work • Pride of workmanship • Low labor turnover

Source: Courtesy of Edward H. Hartmann, International TPM Institute, Inc.

Discuss and get input regarding the status of the equipment with the most important and reliable source that is available to you—the operator. No one is better acquainted with the equipment than the operator who spends many hours every day on the machine and knows every sound, smell, feel, and vibration of the machine. He or she knows exactly when the machine does not quite feel right or is not performing the way it should. The maintenance personnel and engineers are also an invaluable source of information for determining the current status of the equipment.

Of course, there is no substitute for your own senses. Visit the equipment and check the visual condition of the machine. The amount of dirt and grease, any buildup of chips, welding spattering, frayed or loose electrical wires, visible vibration, and misalignments are easily noticeable. It is a simple task to distinguish neglected equipment from equipment that shows pride of ownership.

6.5 OVERALL EQUIPMENT EFFECTIVENESS

A quantitative and a more reliable approach to assessing the current condition of equipment entails observing the equipment over a period of time and calculating its *overall equipment effectiveness* (OEE):

$$OEE = \text{Availability} \times \text{Efficiency} \times \text{Quality}$$

6.5.1   Availability

Availability is also referred to as equipment utilization. It is the ratio of actual operation time to net available time. Net available time is the total available time less any planned downtime, such as breaks or planned preventive maintenance activities. Therefore, if $1\frac{1}{2}$ hours of planned downtime (breaks, PM, meetings, and so on) are scheduled during an 8-hour shift (total available time), then

$$\text{Availability} = \text{Operation time} / \text{Net available time}$$

$$\text{Net available time} = \text{Total available time} - \text{Planned downtime}$$

$$\text{Net available time} = 8 \text{ hours total available time} - 1.5 \text{ hours planned downtime}$$

$$= 6.5 \text{ hours net available time}$$

If 0.5 hour of unplanned downtime occurs, then

$$\text{Operation time} = 8 \text{ hours} - (1.5 \text{ hours planned downtime} - 0.5 \text{ hours unplanned downtime})$$

$$= 6.0 \text{ hours}$$

Therefore

$$\text{Availability} = 6 \text{ hours operation time} / 6.5 \text{ hours net available time}$$

$$= 92.3\%$$

Some of the factors affecting equipment availability that can be measured easily include (but are not limited to) setup and changeovers, adjustments, and software programming, testing, and debugging. Equipment failures also contribute to a decrease in availability.

6.5.2   Efficiency

Efficiency is an indicator of the equipment efficiency and can be calculated as follows:

$$\text{Efficiency} = \text{Theoretical production time} / \text{Actual production time}$$

Actual production time is the same as the operation time from the previous example. Therefore, if the equipment was running (was loaded) for six hours, the actual production time is equal to six hours. How many units should we have produced during this time? The theoretical answer is

6.0 hours $\times$ Theoretical hourly production rate (from time standards)

Assuming the theoretical cycle time is given as 0.0125 hours/unit, theoretical hourly production would be equal to 80 units. Therefore

6 hours $\times$ 80 units per hour = 480 units would be expected

What if only 400 units were produced? The theoretical production time for 400 units = 0.0125 hours per unit $\times$ 400 = 5.0 hours. Hence

$$\text{Efficiency} = \text{5.0 hours Theoretical production time} \, / \\ \text{6.0 hours Production time}$$

$$= 83.3\%$$

Some things that lesser equipment efficiency are stoppages, unavailability of parts at the workstation, and minor adjustments.

6.5.3 Quality

Quality is the ratio of the acceptable parts to the total parts produced. Suppose that during the production run, a total of 400 units were produced, of which 50 were rejected for various reasons, such as scrap or rework.

$$\text{Number of accepted parts} = \text{Total parts produced} - \text{Rejected parts}$$
$$= 400 - 50 = 350 \text{ Acceptable parts}$$

Hence

$$\text{Quality} = 350 \, / \, 400$$
$$= 87.5\%$$

Based on the previous discussion and calculations, our OEE can be calculated as

$$\text{OEE} = \text{Availability} \times \text{Efficiency} \times \text{Quality}$$
$$= 92.3\% \times 83.3\% \times 87.5\%$$
$$= 67.3\%$$

OEE is a measure of overall equipment efficiency, and a higher value indicates an overall better condition than a lower value. Study of each of the factors involved in the calculation of the OEE provides a clue about how to improve the equipment condition and increase its OEE. Figure 6–1 illustrates the formula for OEE calculation and Figure 6–2 presents a numerical example. Figure 6–3 displays major losses that affect overall equipment efficiency.

Overall Equipment Efficiency (OEE)

Availability × Performance × Quality

Availability
(Reflected in operating time loss)

1. Equipment failure loss (all *known* stoppages)
2. Process setup/adjustment losses
3. Startup losses (shift, lunch, repair, and weekend)

Performance
(Reflected in net operating time loss)

1. Idling and minor stoppages (catches *unknown* stops)
2. Speed loss (engineering design speed versus actual speed)

Quality
(Reflected in value operating time loss)

1. Quality defects and rework required

OEE = Availability × Performance × Quality

Note: As TPM and record keeping improve, performance efficiency
will improve at the expense of availability.

FIGURE 6–1 An illustration of the OEE calculation formula and its components

(Source: Courtesy of Subaru-Isuzu Automotive, Inc.)

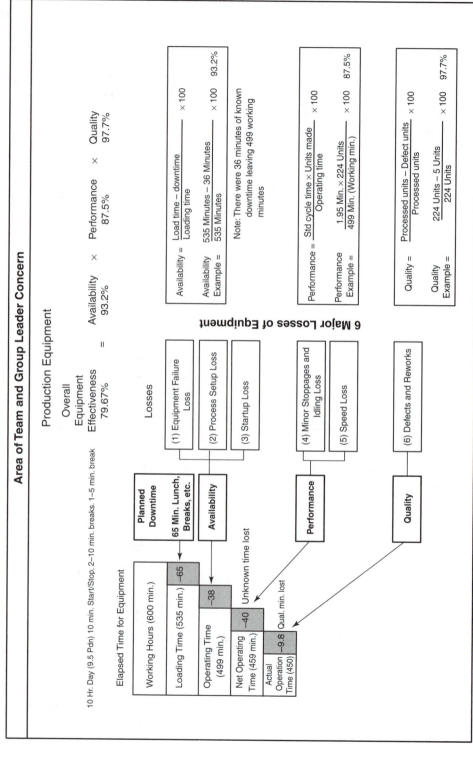

FIGURE 6-2   An illustration of OEE calculation steps

(*Source: Courtesy of Subaru-Isuzu Automotive, Inc.*)

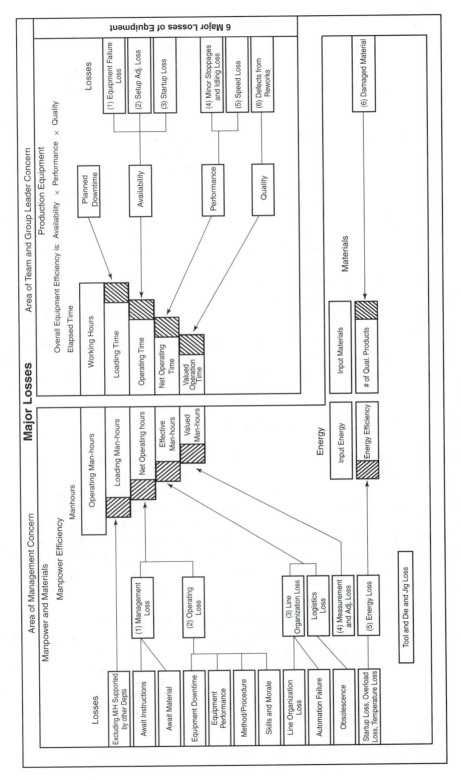

FIGURE 6–3 An illustration of major losses affecting OEE
(*Source: Courtesy of Subaru-Isuzu Automotive, Inc.*)

Along with the assessment of current equipment condition, the current skill levels of the operators also must be determined. Based on the equipment type and condition, the required level of operator skill must be determined and compared with current skill levels. An inventory of current operators' skills can be obtained with the aid of the human resource department and by reviewing personnel files. The inventory should include level of formal education, on-the-job training, workshops, training classes, and seminars attended. Personal interviews can be used to collect these data and more importantly, to determine each individual's level of motivation and interest in facing new opportunities and challenges. Personal interviews also provide the opportunity to add a personal touch and make each operator realize that he or she is a special part of the TPM team and process—because they are. Input and feedback from supervisors also should not be neglected.

Since training is one of the most important factors in implementing TPM, the skills inventory data can help determine the extent, type, and amount of training required to transfer some of the maintenance activities to the operators and create more self-reliance among the operating personnel. To the extent that it is feasible, the information may be used to match certain operator interests with specific and special training.

6.6 TPM ACTIVITIES AND PROCESSES

The heart of the TPM process, simply stated, is the transfer of ownership and responsibility for the basic equipment maintenance activities to trained, well-prepared operators. This is referred to as autonomous maintenance. In this system, production personnel assume an increased responsibility for some of the tasks that have traditionally been performed by the maintenance staff, in order to optimize the performance of their equipment. These responsibilities include maintaining equipment cleanliness, developing lubrication standards, following inspection procedures, and taking the initiative in workstation organization and continuous process improvement.

The purpose of this organized approach in not simply to have operators perform basic maintenance tasks instead of maintenance personnel but to create an environment in which the operator takes an active role in observing and taking actions to improve the overall equipment performance and efficiency.

6.6.1 Cleaning

This most important first step in the TPM implementation process is performed in each production cell or area by a group of individuals. Senior executives, maintenance personnel, and production staff at the Subaru-Isuzu automotive plant in Lafayette, Indiana, speak of their initial cleaning process with such amazing fondness that one would think they were recalling a family reunion or a neighborhood picnic. They recount how at the start of the process *everybody*, including the highest level of management, had put on overalls and goggles and had actively participated in chiseling away weld spattering and wiping away grease and dirt from equipment. To a visitor, the story sounds quite unbelievable. Not that the management had broken down the barriers and taken an active role alongside the operators in the cleaning process, but that the shiny and sparkling equipment had ever been in need of cleaning.

Cleaning also serves as a process to get acquainted with the equipment and discover any hidden defects, cracks, loose bolts, belts, missing parts, broken guards, leaks, and so on. The purpose of the initial cleaning is not to overhaul the equipment but to develop an understanding of the machine. Therefore, it is necessary to obtain all available information about the equipment, such as manuals, drawings, history, and other documentation, to discover important characteristics such as lubrication and oiling points and all safety-related features of the equipment, and to document all new discoveries related to the condition of the equipment.

Cleaning does not end with the initial process. It is an ongoing task to keep the equipment in a like-new condition. Cleaning also plays an important role in identifying contaminants and their sources. Various oils, lubricants, and coolants do not simply appear on the shop floor or the equipment. They leak out from somewhere. Try to identify and document the source, or the root cause, of the problem. The buildup of chips and other process-generated contaminants will result in eventual loss of speed and deterioration in the equipment performance. Identifying the sources of such contaminants can give clues to some simple equipment design changes or modifications, such as curtains and guards that are used to divert and collect the contaminants and minimize their adverse effect on the equipment.

6.6.2 Developing Lubrication Standards

These standards are based on manufacturer recommendations and requirements and lubrication procedures previously performed by maintenance personnel. Allow the modification of these procedures by the operators to best suit their routines. Develop checklists and pictorial images to facilitate communication—drawings and pictures can be used to easily identify and locate lubrication points. Use color codes to identify and match specific lubricants with specific equipment and eliminate the possibility of error and misapplication. Lubrication standards for all equipment should document the type, frequency, and amount of lubricants to be used. Filter specification and change frequency also should be stated. All safety precautions and procedures must be clearly documented. Figure 6–4 displays equipment maintenance standards for a conveyor system. The standards make an effective use of pictorials to reduce or eliminate any chance for confusion or errors.

6.6.3 Inspection

Nobody knows the "feel" of the equipment better than the operator. It is her or his machine. The use of various senses, sight, hearing, touch, and smell, all can communicate with the operator any number of equipment anomalies. General inspections should be part of the process when the operator performs routine cleaning and lubrication procedures. Inspection includes checking the bolts and other fasteners and belts, making adjustments, replacing worn parts, and feeling and checking for hot spots, which may be due to friction, vibration, or other abnormal conditions. Visual inspection of electrical components and connectors can bring to light any unsafe conditions such as loose or damaged wiring. As with the lubrication procedure, development of a checklist greatly facilitates the process. TPM provides the opportunity for the operators to review and modify existing procedures or recommend new preventive and predictive maintenance practices as their involvement and participation with equipment upkeep matures. The operator-initiated inspection also allows for determination and elimination of the root causes of equipment failures.

This aspect of operator involvement also should include training and a working knowledge of the plant computerized maintenance management system (CMMS). The CMMS aids the operator in developing lubrication and inspection checklists and setting the appropriate intervals for each task. Furthermore, since most CMMS packages have the ability to access personnel records to match operator skills with the required craft codes for various PM functions, the CMMS can automatically assign the PM functions to the appropriate operators. Access to CMMS enables the operator to examine and determine the relationship between various maintenance activities and the overall equipment effectiveness and thus provides an excellent opportunity to expand operator involvement in improving and optimizing quality and productivity. This does not reduce or eliminate the need for maintenance staff but increases operators' involvement and contributions. And that is the power of TPM.

6.6.4 Workstation Organization

Housekeeping and workstation organization play an important role in quality and productivity. There should be a place for every item, part, tool, and so on, and every item must be placed in its appropriate location.

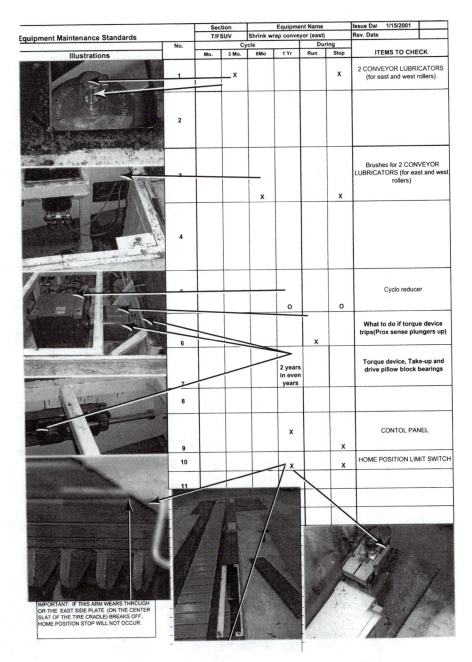

Equipment Maintenance Standards		Section	Equipment Name			Issue Date	1/15/2001	
		T/FSUV	Shrink wrap conveyor (east)			Rev. Date		
	No.	Cycle				During		ITEMS TO CHECK
Illustrations		Mo.	3 Mo.	6Mo	1 Yr	Run	Stop	
	1		X				X	2 CONVEYOR LUBRICATORS (for east and west rollers)
	2							
	3			X			X	Brushes for 2 CONVEYOR LUBRICATORS (for east and west rollers)
	4							
	5			O			O	Cyclo reducer
	6				X			What to do if torque device trips(Prox sense plungers up)
	7			2 years in even years				Torque device, Take-up and drive pillow block bearings
	8							
	9			X			X	CONTOL PANEL
	10			X			X	HOME POSITION LIMIT SWITCH
	11							

IMPORTANT: IF THIS ARM WEARS THROUGH OR THE EAST SIDE PLATE (ON THE CENTER SLAT OF THE TIRE CRADLE) BREAKS OFF, HOME POSITION STOP WILL NOT OCCUR.

FIGURE 6–4 Equipment maintenance standards making use of pictorial displays
(Source: Courtesy of Subaru-Isuzu Automotive, Inc.)

(continued on next page)

Manager	Grp Ldr	Prepared By:		Ref.Number	
		Heichelbech		ITF-886	
HOW TO CHECK		**CRITERIA**		Time	Assign To:
LIFT FLAP AT SOUTH END OF CONV. Brass toggle should be up to lubricate		LIFT SPRING LID & FILLWITH KF-240 OIL		15	PROD.
REMOVE A SLAT OR 2 IN LINE WITH NORTH EDGE OF TAKE-UP PIT		REPLACE THE BRUSHES		60	MAINT.
change the oil		use #150 U		30	MAINT.
Light comes on in panel & conveyor wont run normally		In Manual, run conveyor in reverse approx 6 feet to reset		10	MAINT.
Lubricate with ep-2 rykon grease		Remove lid and inspect, add 3 pumps to fitting		120	MAINT.
VISUAL CHECK		MOTOR RELAY CONTACTS NOT BURNED OR PITTED		15	MAINT
Remove deck plate		Replace the switch arm wire		45	MAINT

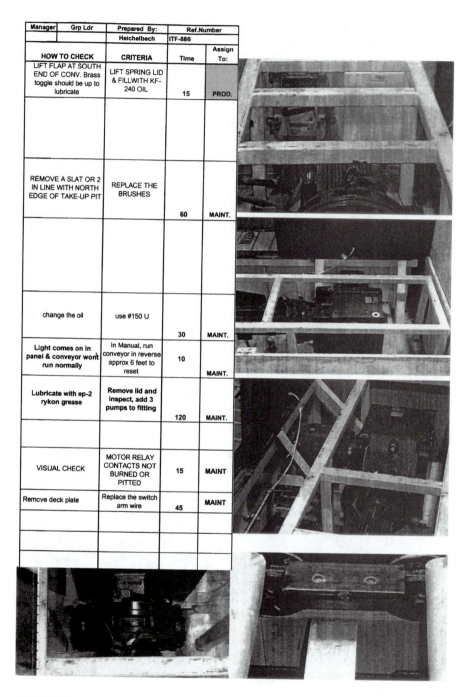

FIGURE 6–4 *(continued)*

Principles of safety, motion economy, and ergonomics must be considered when designing and organizing the workstation. Housekeeping and maintaining workplace organization and cleanliness are an integral part of keeping the equipment and the plant clean and safe.

6.6.5   Continuous Process Improvement

As previously stated, TPM is not merely a maintenance program; rather, it is a comprehensive program aimed at improving overall quality and productivity of the enterprise by increasing performance effectiveness and eliminating waste. Factors such as delays due to material shortage, excessive setup time, manufacturing or material handling steps that are unnecessary, excessive inventory, and a host of other variables have the same adverse effect on productivity and efficiency as do equipment breakdowns and failure. Production personnel are acutely aware of these wasteful procedures and are an invaluable source for process improvement. With proper TPM training and implementation, operators can formally and systematically take an active role in analyzing and improving various manufacturing and processing steps.

The traditional industrial engineering methodology of analyzing steps that are required to perform a task to eliminate, reduce, or simplify each step has an important place in TPM and overall performance improvement. Continuous process improvement (CPI) can and must be used to

- Reduce setup and manufacturing cycle times.
- Reduce and simplify material handling and inspection.
- Eliminate or minimize work in process.

There are a variety of tools such as process charts, process maps, and flow diagrams that are available for this purpose. A small group of participants document all the steps that are currently required to perform a task, as shown in Figure 6–5. This will allow a logical examination of each step to determine how that step contributes to the process and then decide if the step can be eliminated without adversely affecting the integrity of the process. Since operators are most familiar with their own daily and repeated activities, their participation can make significant contributions in reducing setup times, optimizing startup and shutdown procedures, and structuring the handling of various material such as scrap, rework, and so on.

This process also can be used to determine the root cause of various difficulties and to consider methods or procedures to eliminate the root cause of such problems. The use of statistical process control and traditional control charts are invaluable tools to study "trends" and abnormalities in processes and attempt to correct these underlying problems *before* a serious situation occurs.

Process Chart

Part Name: Package Grill

Plant: Shade Tree Grills
Recorded by: RM
Date: 4/3/2004

Summary	Total
○ Operations	46
⇨ Transports	10
□ Inspects	2
D Delays	0
▽ Stores	0
Steps	58
Distance	60 ft

Step #	Operation	Transport	Inspect	Delay	Storage	Description of Method	Method	Distance	Qty Mvd.
1	●	⇨	□	D	▽	grasp retainer	hand		1
2	●	⇨	□	D	▽	prepare retainer	hand		1
3	●	⇨	□	D	▽	place retainer on conveyor	hand		1
4	●	⇨	□	D	▽	grasp grill bottom	hand		1
5	●	⇨	□	D	▽	place grill bottom on retainer	hand		1
6	○	⇨	□	D	▽	retainer moves to next operator	conveyor	6 ft	1
7	●	⇨	□	D	▽	grasp manifold	hand		1
8	●	⇨	□	D	▽	place manifold in grill bottom	hand		1
9	●	⇨	□	D	▽	grasp griddle	hand		1
10	●	⇨	□	D	▽	place griddle in grill bottom	hand		1
11	●	⇨	□	D	▽	grasp heat shield	hand		1
12	●	⇨	□	D	▽	place heat shield in grill bottom	hand		1
13	○	⇨	□	D	▽	retainer moves to next operator	conveyor	6 ft	1
14	●	⇨	□	D	▽	grasp fastener kit	hand		1
15	●	⇨	□	D	▽	place fastener kit in grill bottom	hand		1
16	●	⇨	□	D	▽	grasp wood slats	hand		4
17	●	⇨	□	D	▽	place wood slats in grill bottom	hand		4
18	●	⇨	□	D	▽	grasp plastic component kit	hand		1
19	●	⇨	□	D	▽	place plastic kit in grill bottom	hand		1
20	○	⇨	■	D	▽	retainer moves to next operator	conveyor	6 ft	1
21	○	⇨	□	D	▽	perform visual inspection	x		1
22	●	⇨	□	D	▽	grasp grill top	hand		1
23	●	⇨	□	D	▽	place grill top on grill bottom	hand		1
24	○	⇨	□	D	▽	retainer moves to next operator	conveyor	6 ft	1
25	●	⇨	□	D	▽	grasp leg assemblies	hand		2
26	●	⇨	□	D	▽	place leg assemblies around grill	hand		2
27	○	⇨	□	D	▽	retainer moves to next operator	conveyor	6 ft	1
28	●	⇨	□	D	▽	grasp bottom brace	hand		2
29	●	⇨	□	D	▽	insert bottom brace	hand		2
30	●	⇨	□	D	▽	grasp control panel	hand		1
31	●	⇨	□	D	▽	insert control panel	hand		1
32	●	⇨	□	D	▽	grasp tank holder	hand		1
33	●	⇨	□	D	▽	insert tank holder	hand		1
34	○	⇨	□	D	▽	retainer moves to next operator	conveyor	6 ft	1
35	●	⇨	□	D	▽	grasp axle & wheels	hand		2
36	●	⇨	□	D	▽	insert axle & wheels	hand		2
37	●	⇨	□	D	▽	grasp gas hose	hand		1

FIGURE 6–5   A partial process chart showing steps in a specific task

QUESTIONS

1. What is TPM and what role does it play in the overall productivity of a company?
2. What are some of the fundamental concepts of TPM?
3. How would you relate Deming's concepts of TQM to the culture of TPM?
4. Do you think that TPM can be implemented in a union environment? What are the implications?
5. Use Table 6–1 to classify
   a. your personal automobile
   b. some of the equipment at your place of work
6. Use Table 6–2 to determine the current maintenance productivity of the maintenance department at your place of work.
7. What is OEE and what are the factors affecting it?
8. Why is cleaning considered to be the most important step in equipment maintenance?
9. What is the purpose and importance of developing lubrication standards?
10. What role does CPI (continuous process improvement) play in TPM?
11. What is meant by workplace organization and what is its significance in TPM?

7

TPM IMPLEMENTATION AND PROCESS IMPROVEMENT TOOLS

Overview

Objectives

At the completion of the chapter, students should be able to

- Define and understand benchmarking.
- Understand the five pillars of TPM.
- Understand the eight pillars of maintenance.

- Explain the concept of FMEA.
- Perform a simple FMEA.
- Understand root cause analysis.
- Construct an Ishikawa diagram.
- Construct and perform a simple fault tree analysis.

7.1   INTRODUCTION

A variety of problem-solving and process improvement tools can be used for TPM implementation and continuous quality and process improvement. Although these tools are not specifically designed for TPM, they may prove invaluable for resolving problems, enhancing safety, and improving overall equipment effectiveness. Three of these tools—benchmarking, failure mode and effect analysis, and root cause analysis—are introduced in this chapter.

7.2   BENCHMARKING

It is human nature to learn from others. We learn to apply good practices and avoid practices that tend to be harmful—at least when we use good judgment. In its essence, benchmarking follows the same basic concept. This is not a new idea. Throughout the ages various groups, industrialists, nations, and others have observed and studied the practices of other groups and have attempted to adapt their practices, with varying degrees of success, to their own activities.

Benchmarking can play a significant role in process improvement. Learning from companies that have successfully implemented TPM, gaining wisdom from other practitioners in business and industry about how to improve our processes, and developing the best maintenance programs from the leaders are the basic concepts of benchmarking. Benchmarking is the process of searching for the best in their class and learning from them. It involves identifying companies that are known to have the best practices and that have achieved exceptional results and then attempting to learn and adapt the practices to our own situation and environment. Benchmarking requires an honest self-assessment, an inventory of corporate weaknesses and strengths compared to these leaders, and a determination and a commitment to emulate, not copy, the path that these leaders traveled to get to the top.

Benchmarking requires a company to identify and study the leaders in its own industry or field, rather than settling for second best. But it would be a mistake for a company to limit itself to looking for the best practices only in its own industry, because many processes are applicable across industries. For example, a company can look beyond its own area of production to identify and find best practices in inventory and cycle time reduction, customer service, overall equipment effectiveness, and so on. Then the company must study how these other companies have attained top positions in these areas and how these practices can be adapted to its own industry, culture, and environment.

Some difficulty in benchmarking may arise from the fact that some companies, especially those that might be in direct competition, may be reluctant to share and publicize their business practices. This reluctance come from the propriety nature of the business, confidentiality of the data, or simply from tradition and company policy. It is not a serious obstacle in attaining information on best practices, however. Professional trade shows, seminars, workshops, trade journals and publications, and professional societies are major sources of networking and data.

Keep in mind two very important facts about benchmarking. Benchmarking is not an end or a destination; rather, it is a continuing and ongoing process. That is what is meant by *continuous* improvement. Secondly, the purpose of benchmarking is not to determine *what* the best practices are but *how* these best practices are achieved. In other words, study the process to learn how to apply these practices so that they will become part of the culture of the organization.

7.2.1 A Blueprint for Benchmarking

The blueprint for benchmarking as described by AT&T is explained in the following eight steps:

1. **Conception**: Identify which area(s) of the company may benefit from benchmarking and state the specific need or needs.

2. **Plan**: State the goals and objectives for benchmarking and work out a plan.

3. **Data collection**: You must know your own process. Collect detailed data about your company's current process and practices, as well as about similar processes in industry.

4. **Identifying the best in class**: Identify and select those enterprises that are known to be the best practitioners in that aspect of business.

5. **Collecting data on best practices**: Collect detailed data on the practices of those companies that have been identified as best in class. Collect data on how these companies achieved that status.

6. **Comparison**: Compare and evaluate your company's practices and methods with the practices of those who are best in class.

7. **Implementation**: Develop and implement plans to incorporate these practices into your own operations and procedures.

8. **Evaluation**: Assess your success. Measure your achievements and compare them with your goals and objectives. Determine where the shortfalls are and what needs to be done in order to get back on track.

7.3   THE FIVE PILLARS OF TPM

TPM is rooted in five specific and clearly defined solid principles. These underlying concepts or principles, which are the fundamental doctrine of TPM, can be referred to as the five pillars of TPM. Figure 7–1 shows a conceptual portrayal of TPM and its supporting pillars. The principle depicted on each pillar is briefly explained in the following paragraphs.

7.3.1   TPM Principles

1. **Planned maintenance**. All maintenance activities should be *planned* rather than reactive, which will result in greater equipment reliability and uptime, reduced quality defects, and improved safety.

2. **Maintenance free**. This system of thinking and practice at the concept and design stage helps create equipment that requires less maintenance, and when maintenance is required, the equipment can be brought back into operation in less time.

3. **Individual *kaizen***. *Kaizen* is considered the single most important concept in Japanese management philosophy; it is a constant search for improvement. *Kaizen* encompasses any activity that is directed towards improvement; therefore, any team activity that focuses on increasing equipment effectiveness, quality improvements, or any other areas of business such as safety, customer service, cost reduction, waste elimination, employee education, and productivity falls in the circle of *kaizen*.

4. **Education and training**. One of the most important concepts in TPM is education and training. Effectiveness of employees directly depends on their knowledge of the principles of TPM, and their understanding of the equipment and processes.

5. **Self-maintenance**. Refers to a basic program of maintenance activities that are performed by the operator. An operator self-maintenance program includes, but is not limited to, initial cleaning of equipment; devising countermeasures (cause and effect study) to

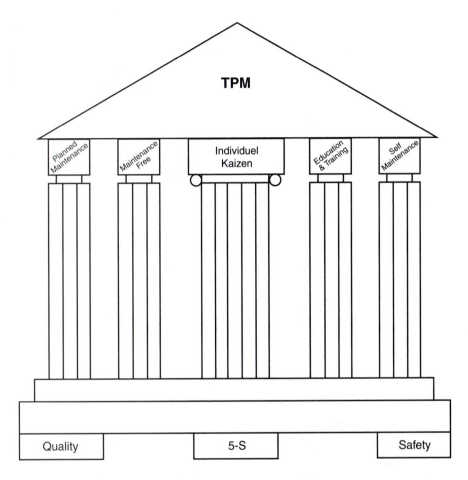

FIGURE 7–1   A conceptual depiction of TPM and its five fundamental pillars
(Source: Courtesy of Subaru-Isuzu Automotive, Inc.)

eliminate or reduce dust and dirt; routine cleaning, inspection, and lubrication; and total proactive management focused on safety, quality, and cost.

Each pillar is an important fundamental principle, but their strong interrelationship should not be lost to the reader. Many aspects of these pillars are interwoven and have common ground.

It is interesting to note that Figure 7–1 shows the pillars themselves resting on the concepts of quality, safety, and the 5-S principle. The 5-S principle is a concept used to describe what proper housekeeping and organization means. The 5-S refers to five Japanese words that start with the letter "s," and they convey the concepts of tidiness, organization, cleanliness, and discipline. Specifically, the 5 Ss describe the Japanese philosophy of workplace organization and refer to *seiri* (sorting, arrangement), *seiton* (organization, proper housekeeping), *seiso* (sweeping, workplace hygiene), *seiketsu* (neatness, cleaning), and *shitsuke* (strictness or discipline).

7.4   THE EIGHT PILLARS OF MAINTENANCE

Based on the five pillars of TPM (keep in mind that TPM is an overall organization concept, not merely a maintenance program), the maintenance activities under the TPM umbrella are built on the eight pillars of maintenance. These points have been discussed in varying amount of detail in other sections of this text; therefore, they are only given a cursory treatment here. Figure 7–2 display a sample master plan for implementing TPM and its eight pillars of maintenance. The eight pillars of maintenance are as follows:

1. **Support and guidance for operator self-maintenance activities**. This emphasizes management's role and responsibility in providing education and training programs to enhance operators' skills and participation.

2. **Reduction of breakdowns**. Initiate activities to achieve zero failures. The objective is to reduce failures for the entire plant. Identify problem areas and determine root causes. Apply solutions to eliminate failures and track these measure to confirm their effectiveness. These measures should extend mean time between failure (MTBF) and mean time to repair (MTTR).

3. **Planned maintenance activities.** Review the existing preventive maintenance program. Check equipment lists for completeness and accuracy and prioritize relative productivity, safety, and cost. Devise the best type of maintenance program (time based, condition based, overhaul, and so on) for each.

4. **Lubrication management.** Set up programs for lubrication. Determine type, how, when, and where. Eliminate leaks and failure due to lack of lubrication.

5. **Spare part management.** Manage spare part inventory for planned, not breakdown, maintenance. Determine what is needed and what is excess inventory. Develop standards for inventory items, their quantity, reorder point, and a control system.

Master Plan

8 PILLARS OF MAINTENANCE ACTIVITIES	Section Activities		2000	2001	2002

BASIC TPM SCHEDULE

Pillar #1 — Support & Guidance for Operator Self-Maintenance Activities
- JIPM Consistency Award
- Consulting Visits
- Support TPM Operator Self-Maintenance Program
- Indicate Support and Targets for Operator F-Tag Closure
- Program for Transfer and Training of PM Items to Operator Self-Maintenance Standards

Step 6.1 (A-B-C-D) — Continuous support of Production on the implementation and growth of step 5 activities. Develop a * trained to TPM standard * list. Promote the vista & kaizen program to achieve lasting improvements.
Step 6.3 (A-B-C-D) / Step 7 Operator Self-Maintenance

Pillar #2 — Zero Failure Activities
- Indicate Targets for:
  - Downtime Failure Reduction
  - Frequency Failure Reduction
  - Process for Reduction
  - Failure Mode Analysis - (Biggest Losses)
  - Downtime Report and 5-Why Training for all Maintenance Associates
  - Track Re-Occurrence of problems

Develop plans and evaluate drawings and installation of new equipment to improve quality, reduce start-up problems and enhance productivity. Maintenance to track 'Worst Six' and insure no problems. Analyze the problem and implement the appropriate countermeasures. Reduce to "ZERO"

	Bench Mark	Year and Target	Year and Target	Year and Target
Downtime Failure Reduction	0.53%	0.34%	0.31%	0.27%
Frequency Failure Reduction	12/month	8/month	7/month	6/month
Failure Mode Analysis	3 per month	1 per month	6 per year	3 per year
Track Re-Occurrence	16%	64%	88%	100%

Pillar #3 — Establish Planned Maintenance System
- Continuous Improvement of Preventative Maintenance Process
  - PM Program, PM Calendar, Standards and Checklists
- Target for PM's Completed vs. Scheduled
- Plan and Implement Long Cycle PM Program by (9/30/2000)

Analyze — Review / Continue Implementation
- Brake/motor replacement program: from CMS to FB series
- Remain current on monthly PM activities
- Continuous improvement of PM schedule. Find the wear items and replace them prior to failure. Extend the life by proper fit, adjustment & material selection.
- Install lifter chain stretch measurement. Replace TO-906 chain. Replace 3 DIL chains.

	Bench Mark			
PM's Completed	95%	97%	97%	98%

Pillar #4 — Lubrication Management Activities
- Lubrication Management Efficiency Improvement Schedule
  - All Lubes scheduled, Over use of lubs minimized, Types and kinds to minimize,
  - Color Code Chart
  - Leak Control Program

Confirm autonomous lubrication activities. Specify and Insure proper lubrication type and quantity is applied. Identify * Permanently Lubricated * bearings and develop replacement program before failure. Run entire year without bearing, roller or lubrication failure.

Pillar #5 — Spare Parts Management
- Indicate schedule for spare parts program efficiency improvement
- Schedule of Section Spare Parts Management System Standards

Investigate shared inventory accessibility. Develop bar code parts system using SAP info record numbers. Selection of inventory management software. Develop systematic purchase program. Implement visual controls.

Pillar #6 — Maintenance Cost Management
- Indicate Target for overall budget Reduction
- Target for Outside Contractors Cost Reduction
- Schedule and Target for Cost Savings Kaizens

Expand cost tracking charts to include Main Hours Per Unit, Parts Cost Per Unit, and budget allocation. Maximize cost effective 'in house' rebuild. Continue cost effective in house rebuild.

	Bench Mark			
	15%			

Pillar #7 — Maintenance Efficiency Improvement
- Identify Targets for:
  - Mean Time Between Failure (MTBF)
  - Mean Time To Repair (MTTR)
  - Fail Frequency Ratio (FFR)
  - Failure Duration Ratio (FDR)
- Targets for Breakdown Maintenance vs. Routine Maintenance Plans and Targets for Predictive Maintenance Technique Implementation Maintenance Prevention (FREE) Activities

Utilize MTBF and MTTR tracking to measure efficiency. Confirm new equipment coming in for model change meets SIA specification, reducing need to modify it to achieve reliability. Evaluate and eliminate out source venders where possible. Continual evaluation of time base PM's to chage to condition based.

	Bench Mark	Target		
MTBF	1292	1750	1825	1900
MTTR	6.38	4.1	3.6	3.1
FFR	0.06	0.04	0.04	0.03
FDR	0.53%	0.34%	0.31%	0.27%

Pillar #8 — Maintenance Training
- Identify Training level Targets for Maintenance Skill Training
  - Internal
  - External
- 1* understand 2* can do with help 3* can do alone 4* can teach others

Skill assessment and traing level self evaluation. Select, target training in the training center as they can fit into work schedule.

	Bench Mark			
Internal	2.4	3	3.2	3.3
External	3	3.2	3.2	3.3

Pre-Audit, Final Audit

FIGURE 7-2 A sample master plan for implementing the eight pillars of maintenance
(*Source: Courtesy of Subaru-Isuzu Automotive, Inc.*)

179

6. **Maintenance cost management.** Establish a maintenance cost management system to track and control the overall budget, including inventory costs, direct and indirect supplies, personnel hours, and contractors' expense. Initiate cost reduction measures.

7. **Maintenance efficiency improvement.** Analyze the current maintenance system. Research the elements and principles of TPM. Compare and evaluate time-based, condition-based, preventive, corrective, and predictive maintenance with respect to each piece of equipment and select the most feasible practice for each. Low priority items that do not affect productivity may be candidates for corrective maintenance.

8. **Training.** Enhance training of all operators. Identify needs and plan for long-term improvement of their education and skills.

7.5 FAILURE MODE AND EFFECT ANALYSIS

As the responsibility of operation employees for equipment improvement increases and their level of training and sophistication expands, the use of additional predictive and prevention tools may be warranted. One such procedure involves understanding potential failures and their consequences on the equipment and the overall operation system. Another tool is used to determine the underlying cause of failures so that the root causes, not the symptoms, are corrected. The former is referred to as failure mode and effect analysis (FMEA) and will be discussed in the following paragraphs. The latter, root cause analysis, is discussed in the next section.

   This is not intended as an in-depth discussion of FMEA and the purpose is not to develop experts in the discipline; interested readers are referred to a substantial body of available literature on the subject. The goal here is to provide a basic understanding of the subject and develop an appreciation for its use, albeit a cursory one, in predicting potential equipment failures and the subsequent effects of such failures.

   The purpose of FMEA is to identify *potential* failures, that is, all the ways in which equipment, a process, or a product can fail; to determine how these failures can affect the overall operation of the equipment or the entire system; and to propose solutions or a course of action to eliminate these potential failures.

7.5.1 FMEA Categories

Traditional FMEA is divided into four general types including design, process, application, and service FMEA. Design or product FMEA focuses on the potential failure modes due to engineering design blunders or inadequacies. The best time to attempt this analysis is during the design stages of the equipment. Involving maintenance and operation personnel during the design process can eliminate a host of future and potential failures. The concept of *design for maintainability* is concerned with building into the equipment (product) those features that eliminate, reduce, and simplify maintenance needs and procedures. FMEA steps at the design stage also aid in identifying weak links and areas in which equipment reliability can be improved. Product FMEA also is useful for identifying sources of potential failures with existing equipment, which may warrant design modification on the current or the future generation of machines.

Process FMEA attempts to identify the steps in the procedures and processes required in performing the task that may cause a failure and assess the effect of these failures. Once again, the initial equipment or product design and its strengths or weaknesses have a significant bearing on the eventual processes and their propensity for failure.

Identifying the potential failures or risks associated with the use or application of the end product, those risks associated with the customer, is the concern of application FMEA. This type of FMEA assesses the potential for failures while the product is used by the customer and how the failures may affect the user. Application FMEA attempts to devise a solution to avert such failures.

The fourth category of FMEA is service FMEA. Service FMEA is an attempt to identify potential failure modes of a service activity, their consequences, and then to provide subsequent corrective actions. Consider maintenance as a service activity. What are the potential failures of a maintenance activity? The focus of the analysis is to predict how the service, in this instance the maintenance task, could fail to fulfill its intended function, investigate the consequences of such failure, and select a remedy and corrective actions in order to avert the possibility of such failures. A vast array of maintenance-related functions can be subjected to FMEA and may benefit immensely from it. This exercise will enable us to identify the effect of potential failures and provide the opportunity to proactively devise a means for dealing with or averting such failures. One simple example of a maintenance activity is lubrication. What are the potential failures associated with this activity? One might be the failure to lubricate at all. Another might be overlubrication or the use of the wrong grade of lubricants. Each one of these improper actions indicates a potential failure. What are the results (the effects) of each of these *failures* on performing the function correctly? Based on the severity of the effects of the failures, proper procedures are put in place to avoid these potential failures.

Utility companies, financial institutions, hospitals, government agencies, and educational institutions also have processes that will benefit from service FMEA. The goals of service FMEA are to develop solutions to potential failures so that customers' (internal and external) expectations about quality, reliability, and productivity are met while controlling the cost.

FMEA as a discipline has its roots in the aerospace industries in the 1960s, especially the Apollo program. It has since been recognized for its potentials for detecting, assessing, and avoiding risks by various industries and organizations. Among the proponents of FMEA are the U.S. Department of Defense, the automotive industry, the U.S. Food and Drug Administration (FDA), and the Federal Aviation Administration, to name a few. In 1991, ISO 9000 recommended the use of FMEA, and application of FMEA became a requirement for QS 9000 certification in 1994.

7.5.2   Performing a Service FMEA

An appropriate form is necessary to perform a service FMEA. No universally accepted or standard service FMEA form exists, however. Various organizations adopt or design their own form to reflect their needs and those of their customers. Figure 7–3 is a sample service FMEA form and the following paragraphs explain the various fields on the form.

The header items are used to identify the particular service function and the person or the group responsible for providing that service and the date of such service. The provider may be the operator, the maintenance department, or a service contractor. The remaining header items are used to identify the FMEA team or group and the dates on which such analysis was performed.

Service   In the column, write an exact description of the service or the task as it currently is, not as it should or could be, in clear and concise terms. The information in this column should explain what the function or the purpose of the service is. Examples may include the following entries:

- Check all electrical connectors
- Check the level of coolants in reservoirs
- Answer all incoming calls before the third ring

Potential failure mode   For each service stated, potential failure mode(s) must be listed. There may be more that one failure mode for each service function. Clear and concise statement of the failure mode greatly facilitates the identification of effect and causes of such failures.

Service name: _____   Prepared by: _____

Service responsibility: _____   FMEA date: _____

Service date: _____   FMEA rev. date: _____

Page ____ of ____ pages

Service	Potential failure mode	Potential effect(s) of failure	Potential cause(s) of failure	OCC	SEV	DET	RPN	Recommended action(s)	Individual/area responsible and completion date	Action taken

Team signatures

_____

_____

_____

_____

Approval signatures

_____

_____

_____

_____

FIGURE 7–3   A sample form for conducting a service FMEA

To clearly understand what "potential failure mode" means, think of what would happen if this service were not performed or were performed in an incorrect way. Potential failure modes as a result of improperly performing the services listed previously (failing to provide the service as specified) could include the following:

- Missed loose electrical connections
- Missed frayed wires
- Low coolant in the reservoir
- Annoyed customer

Potential effects of failure   The entries in this column on the service FMEA form attempt to predict the effect of the failure of the service, or the consequences of the failure. What are the consequences or how are the customer, the company, the environment, and so on affected if this failure occurs? Historical data and past experiences are a good source of information to answer these questions and provide the entries for this column. Some of the answers are obvious. When a utility company fails to provide service, resulting in a blackout in the heat of summer, it is not difficult to imagine the potential effects of such a failure. What are the potential effects in the case of the failure modes listed previously?

- Loss of power
- Fire
- Electrical shock, injury, and death
- Loss of customer

Potential causes of failure   This category is probably the most important aspect of the analysis. Determining what caused the failure is necessary so that it can be eliminated and its reoccurrence can be avoided. The emphasis is to identify the *cause*, or the *root cause*, of the problem and *not* the symptom. When an oil seal is damaged and causes an oil leak, replacing the seal is simply dealing with the symptom. If it can be determined that, for example, improper tolerances and/or specifications caused internal friction between the parts and the metal shavings, which in turn damaged the seal and resulted in the oil leak, then the root cause of the problem has been found. Keep in mind that there may be more than one cause for a potential failure mode. In our example, what might have caused our failure modes?

- Inadequate instruction
- Insufficient training

TABLE 7–1

Suggested FMEA Occurrence Evaluation Criteria

Probability	Likely failure rates	Ranking
Very High: Persistent Failures	≥ 100 per thousand pieces	10
	50 per thousand pieces	9
High: Frequent Failures	20 per thousand pieces	8
	10 per thousand pieces	7
Moderate: Occasional Failures	5 per thousand pieces	6
	2 per thousand pieces	5
	1 per thousand pieces	4
Low: Relatively Few Failures	0.5 per thousand pieces	3
	0.1 per thousand pieces	2
Remote: Failure Is Unlikely	≤ 0.01 per thousand pieces	1

Source: Copyright © 1993, 1995, 2001; DaimlerChrysler Corporation, Ford Motor Company, General Motors Corporation.

Occurrence (OCC)   The occurrence column provides information about the expected frequency of each failure as a result of each given cause. Reliability studies may provide a mathematical basis for this information. Data from similar services or historical data also can provide adequate information. Occurrence can be assigned a numerical ranking from 1 to 10. Based on QS 9000 guidelines, a ranking of 10 is assigned if the probability of failure is "very high" or in the case of persistent failures. A ranking of 4 to 6 is assigned if the probability of failure can be estimated as "moderate." If the probability of failure is "remote" or the possible failure rate is considered to be less than 1 in 100,000, then a ranking of 1 is used. Other values are used as the evaluation team ranks the probability of failure between very high and remote. A complete list of suggested FMEA occurrence evaluation criteria is presented in Table 7–1.

Severity (SEV)   Recall the *effect* of the potential failure. Loss of customers, varying degree of annoyance, injury, or death can occur as the result of failure. The severity column is used to rate the seriousness of the effect of the failure on the system, customer, and/or the environment. Severity can be categorized based on a criticality rating. The four categories are as follows:

- Criticality 1: Such failures may result in loss of human life, serious injury, significant loss of function, or a combination of such catastrophes.

- Criticality 2: A failure may result in a significant loss of function or a major portion of the facility. These failures are critical.

- Criticality 3: Such failures may result in minor damages. These failures are considered to be minor.

- Criticality 4: Failures in this category are insignificant in nature and do not impact the system. They tend to be more of a nuisance and an annoyance.

A severity ranking methodology consistent with the ranking scheme used for occurrence ranking may also be used. Based on the QS 9000 guidelines, a ranking of 1 through 10 is used, with 1 denoting "no discernible effect" and 10 indicating a "hazardous condition without warning," or a severity with criticality of 1. A moderate severity may be given a rank of 6. Other numerical values are assigned based on the discretion and the judgment of the FMEA team. Table 7–2 shows complete FMEA severity evaluation criteria. The criteria can be used to judge the effects of the potential failure mode on either the customer or the manufacturing/assembly operations.

Detection (DET)   Detection refers to the ability to detect the failure before it can adversely affect the system, or in the case of the service FMEA, before the failure can reach the customer. Therefore, service shortfalls and inadequacies should be identified as early as possible. Historical data and deficiencies of similar services and their past failures can provide valuable data. Brainstorming efforts by the FMEA team should concentrate on determining how and how quickly the failure can be discovered and recognized. Understanding system reliability, experimentation, and simulation can provide clues about the probability and the nature of the failure, which will aid failure detection. A 10-point detection rating system similar to the previous two rankings can be used to rank the likelihood or the probability of detecting a failure, assuming that the failure *has* occurred, before the failure can adversely affect the system or reach the customer. According to this rating criteria, if the detection is "almost impossible," an absolute certainty of nondetection exists, that is to say, if the existing control systems cannot or do not detect failures or no failure detection control system exists, a rank of 10 is assigned. On the opposite side of the scale, if the likelihood of detection is "very high," or if the existing controls most likely will detect a failure mode, a rank of 1 is given. A "moderate" chance of detection will get a rating of 5. Once again, the team uses its discretion to assign in-between values. A complete FMEA detection evaluation criteria is shown in Table 7–3.

TABLE 7–2
Suggested FMEA Severity Evaluation Criteria

	Criteria: Severity of Effect	Criteria: Severity of Effect	
	This ranking results when a potential failure mode results in a final customer and/or a manufacturing/ assembly plant defect. The final customer should always be considered first. If both occur, use the higher of the two severities.	This ranking results when a potential failure mode results in a final customer and/or a manufacturing/ assembly plant defect. The final customer should always be considered first. If both occur, use the higher of the two severities.	
Effect	Customer effect	Manufacturing/ assembly effect	Ranking
Hazardous Without Warning	Very high severity ranking when a potential failure mode affects safe vehicle operation and/or involves noncompliance with government regulation without warning.	Or may endanger operator (machine or assembly) without warning.	10
Hazardous With Warning	Very high severity ranking when a potential failure mode affects safe vehicle operation and/or involves noncompliance with government regulation without warning.	Or may endanger operator (machine or assembly) with warning.	9
Very High	Vehicle/item inoperable (loss of primary function).	Or 100% of product may have to be scrapped, or vehicle/item repaired in repair department with a repair time greater than one hour.	8
High	Vehicle/item operable but at a reduced level of performance. Customer very dissatisfied.	Or product may have to be sorted and a portion (less than 100%) scrapped, or vehicle/item repaired in repair department with a repair time between a half hour and an hour.	7

(continued on next page)

TABLE 7–2

(continued)

Effect	Customer effect	Manufacturing/ assembly effect	Ranking
Moderate	Vehicle/item operable but comfort/convenience item(s) inoperable. Customer dissatisfied.	Or a portion (less than 100%) of the product may have to be be scrapped with no sorting, or vehicle/item repaired in repair department with a repair time less than a half hour.	6
Low	Vehicle/item operable but comfort/convenience item(s) operable at a reduced level of performance.	Or 100% of product may have to be reworked, or vehicle/ item repaired offline but does not have go to repair department.	5
Very Low	Fit and finish/squeak and rattle item does not conform. Defect noticed by 50% of customers.	Or the product may have to be sorted, with no scrap, and a portion (less than 100%) reworked.	4
Minor	Fit and finish/squeak and rattle item does not conform. Defect noticed by 50% of customers.	Or a portion (less than 100%) of the product may have to be reworked, with no scrap, online but out-of-station.	3
Very Minor	Fit and finish/squeak and rattle item does not conform. Defect noticed by discriminating customers (less than 25%).	Or a portion (less than 100%) of the product may have to be reworked, with no scrap, online but in-station.	2
None	No discernible effect.	Or slight inconvenience to operation or operator, or no effect.	1

Source: Copyright © 1993, 1995, 2001; DaimlerChrysler Corporation, Ford Motor Company, General Motors Corporation.

TABLE 7–3

Suggested FMEA Detection Evaluation Criteria

Detection	Criteria	Inspection Types			Suggested range of detection methods	Ranking
		A	B	C		
Almost Impossible	Absolute certainty of nondetection.			X	Cannot detect or is not checked.	10
Very Remote	Controls will probably not detect.			X	Control is achieved with indirect or random checks only.	9
Remote	Controls have poor chance of detection.			X	Control is achieved with visual inspection only.	8
Very Low	Controls have poor chance of detection.			X	Control is achieved with double visual inspection only.	7
Low	Controls may detect.		X	X	Control is achieved with charting methods, such as SPC (Statistical Process Control).	6
Moderate	Controls may detect.		X		Control is based on variable gauging after parts have left the station, or Go/No Go gauging performed on 100% of the parts after parts have left the station.	5
Moderately High	Controls have a good chance to detect.	X	X		Error detection in subsequent operations, or gauging performed on setup and first-piece check (for setup causes only).	4
High	Controls have a good chance to detect.	X	X		Error detection in-station, or error detection in subsequent operations by multiple layers of acceptance: supply, select, install, verify. Cannot accept discrepant part.	3
Very High	Controls almost certain to detect.	X	X		Error detection in-station (automatic gauging with automatic stop feature). Cannot pass discrepant part.	2
Very High	Controls certain to detect.	X			Discrepant parts cannot be made because item has been error-proofed by process/product design.	1

Inspection Types:
  A. Error-proofed
  B. Gauging
  C. Manual Inspection

Source: Copyright © 1993, 1995, 2001; DaimlerChrysler Corporation, Ford Motor Company, General Motors Corporation.

Risk priority number (RPN)   RPN is calculated by multiplying the three rating values of occurrence, severity, and detection. RPN values are used to rank order or prioritize potential failures; therefore, a RPN by itself has no particular meaning. The higher the RPN, the higher is the priority for the corresponding potential failure.

Recommended action   Unless specific recommendations and a course of action are stated, the FMEA is an exercise in futility. The objectives of FMEA are to investigate and uncover potential failure modes, determine their effect, and recommend a course of action to avert the failure. FMEA team must prioritize potential failures based on their calculated RPN and determine appropriate remedies in order to reduce or completely eliminate service deficiencies and their resultant failure modes.

Action taken   This section is for follow-up purposes. A detailed description and explanation of the corrective actions recommended by the FMEA team should be provided to ensure that follow-up actions were taken to remedy and eliminate potential failure modes. If the follow-up actions are not recorded, the FMEA and the resulting recommendations will have served no useful purpose.

FMEA is a dynamic process, as is any manufacturing or service environment in which the FMEA activity is performed. It is a process that does not remain constant and static. The business environment, the processes, the customers, and the operation environment are constantly changing. So must, and does, the FMEA methodology.

Once an FMEA activity has been conducted and corrective actions have been taken, further studies are necessary to evaluate the effectiveness of the previous recommendations and the subsequent actions. In the first place, it is important to determine if the recommendations were successfully carried out. And if so, what was the effect of these actions? Has the potential for the failure mode been eliminated? If not, why not? Have the recommendations been effective in lowering the severity and the occurrences of the failure mode or increased the probability of its detection if the failure mode occurs, or overall has the value of the RPN been lowered? Follow-up studies and investigations should answer these and other related questions and provide further recommendations for improving the quality and reliability of the service. Figure 7–4a identifies and graphically illustrates the process sequence for performing FMEA. In Figure 7–4b a sample completed FMEA is presented. The completed sample form also shows the recommended actions and the results of the actions taken. A blank form is displayed in Figure 7–4c.

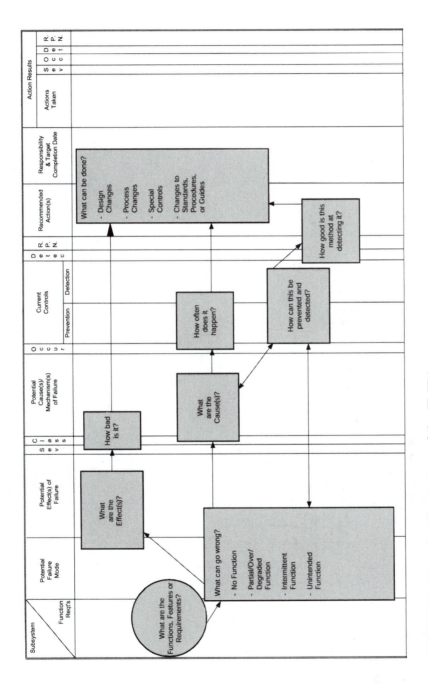

FIGURE 7–4a An illustration of the FMEA process sequence

(*Source: Copyright © 1993, 1995, 2001; DaimlerChrysler Corporation, Ford Motor Company, General Motors Corporation.*)

Item Front Door L.H./H8HX-000-A Process Responsibility Body Engrg. Prepared By J. Ford - X6521 - Assy Ops

Model Years(s)/Vehicle(s) 199X/Lion 4dr/Wagon Key Date 9X 03 01 ER FMEA Date (Orig) 9X 05 17 (Rev) 9X 11 06

Core Team A. Tate Body Engrg, J. Smith-OC, R. James-Production, J. Jones-Maintenance

Process Function Requirements	Potential Failure Mode	Potential Effect(s) of Failure	C l a s s / S e v	Potential Cause(s)/Mechanism(s) of Failure	O c c u r	Current Process Controls Prevention	Current Process Controls Detection	D e t e c	R. P. N.	Recommended Action(s)	Responsibility & Target Completion Date	Action Results Actions Taken	S e v	O c c	D e t	R. P. N.
Manual application of wax inside door	Insufficient wax coverage over specified surface	Deteriorated life of door leading to: • Unsatisfactory appearance due to rust through paint over time • Impaired function of interior door hardware	7	Manually inserted spray head not inserted far enough	8		Visual check each hour-1/shift for film thickness (depth meter) and coverage	5	280	Add positive depth stop to sprayer	MFG Engrg 9X 10 15	Stop added, sprayer checked on line	7	2	5	70
										Automate spraying	Mfg Engrg 9X 12 15	Rejected due to complexity of different doors on same line				
To cover inner door, lower surfaces at minimum wax thickness to retard corrosion			7	Spray heads clogged - Viscosity too high - Temperature too low - Pressure too low	5	Test spray pattern at start-up and after idle periods, and preventive maintenance program to clean heads	Visual check each hour-1/shift for film thickness (depth meter) and coverage	5	175	Use Design of Experiments (DOE) on viscosity vs. temperature vs. pressure	Mfg Engrg 9X 10 01	Temp and press limits were determined and limit controls have been installed - control charts show process is in control Cpk=1.85	7	1	5	35
			7	Spray head deformed due to impact	2	Preventive maintenance programs to maintain heads	Visual check each hour-1/shift for film thickness (depth meter) and coverage	5	70	None						
			7	Spray time insufficient	8		Operator instructions and lot sampling (10 doors / shift) to check for coverage of critical areas	7	392	Install spray timer	Maintenance 9X 09 15	Automatic spray timer installed - operator starts spray, timer controls shut-off control charts show process is in control Cpk=2.05	7	1	7	49

SAMPLE

FIGURE 7–4b A sample FMEA and actions

(*Source: Copyright © 1993, 1995, 2001; DaimlerChrysler Corporation, Ford Motor Company, General Motors Corporation.*)

POTENTIAL
FAILURE MODE AND EFFECTS ANALYSIS
(PROCESS FMEA)

FMEA Number _____

Page _____ of _____

Item _____   Process Responsibility _____   Prepared By _____

Model Year(s)/Vehicle(s) _____   Key Date _____   FMEA Date (Orig.) _____ (Rev.) _____

Core Team _____

Process Function Requirements	Potential Failure Mode	Potential Effect(s) of Failure	S e v	C l a s s	Potential Cause(s)/ Mechanism(s) of Failure	O c c u r	Current Design Controls - Prevention - Detection	D e t e c	R. P. N.	Recommended Action(s)	Responsibility & Target Completion Date	Action Results				
												Actions Taken	S e v	O c c	D e t e c	R. P. N.

FIGURE 7–4c   A sample FMEA form

(*Source: Copyright © 1993, 1995, 2001; DaimlerChrysler Corporation, Ford Motor Company, General Motors Corporation.*)

193

A SHOCKING FMEA

Subaru-Isuzu Automotive, Inc. (SIA), located in Lafayette, Indiana, is a world-class model for TPM. The plant was established in 1987, spans 800 acres, and employs more than 3,000 employees. The facilities operate with two production shifts producing nearly one-quarter of a million passenger cars and SUVs per year. SIA holds ISO 9002 certification and has been a successful TPM facility for more than 10 years. SIA received the TPM Award of Excellence in 1998.

SIA has based its outstanding TPM program on an aggressive employee education and training program designed to enhance the problem-solving abilities and technical skills of their associates. SIA's investments in such educational programs and the employees' involvement and participation in quality and productivity improvements are exemplary.

Although SIA has many ongoing quality and productivity activities worth looking at more carefully, in this case study, we will examine an FMEA activity, shown in Figure 7–5, that was carried out to study a potential electrical hazard and find a solution to eliminate the hazard.

The header information provides the required documentation with information about the location, date, and so on. The problem classification section is used to specify the nature of the problem as it relates to quality, safety, and so on.

A clear statement describes the potential hazard. Although this problem has not yet occurred, a proactive study of the situation reveals that the design of the connector provides the potential for hazard. Due to the shape of the connector and the way it is grasped while disconnecting the plug, the wires are subjected to extreme strain, which will eventually result in their fraying and breaking, creating a potential failure mode. The accompanying photographs in Figure 7–5 clearly show the problem. Illustrations such as these enhance the understanding of the problem statement and aid in visualizing possible solutions.

The subsequent effects of such failures, in addition to equipment downtime, may include electrical shocks and burns to employees, risk of fire, and further damage to the equipment and the facilities.

As a result of the study, countermeasures were suggested and implemented. The FMEA form clearly states how these countermeasures will avert the potential problem. Hence, overall equipment efficiency and productivity are improved.

As mentioned earlier, there are no standard forms for performing FMEA. It is important to note that, although this study followed a simpler process than suggested in the discussion of FMEA in the chapter text the results of the study are just as valid and significant in averting a potential failure mode and improving uptime.

	Control # tfmaint-
Date written: 1/31/2001	
Equipment Name: Power seat checker	Maint. G/L: Jim H
Area / Location: Trim sta. 49	Written By: Jim H

Problem Classification: ☐ Quality   ☐ Safety   ☐ Ergonomics   ☐ Maintenance   ☐ Operator   ☑ Durability

Other: _____

Problem Description:
The connector supplied is made without strain relief, using the wires to disconnect is likely, leading the the wires breaking

Problem Details:

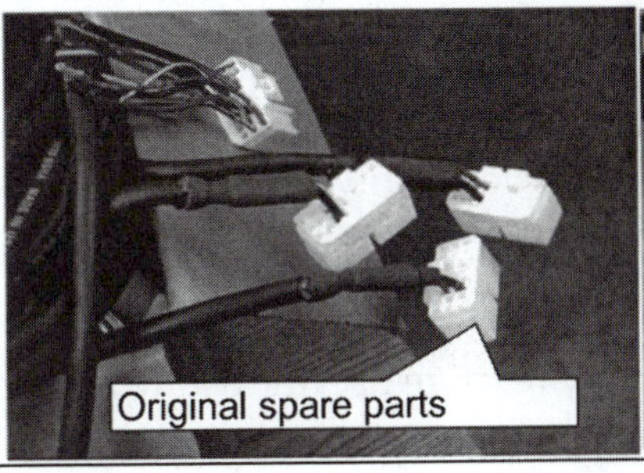

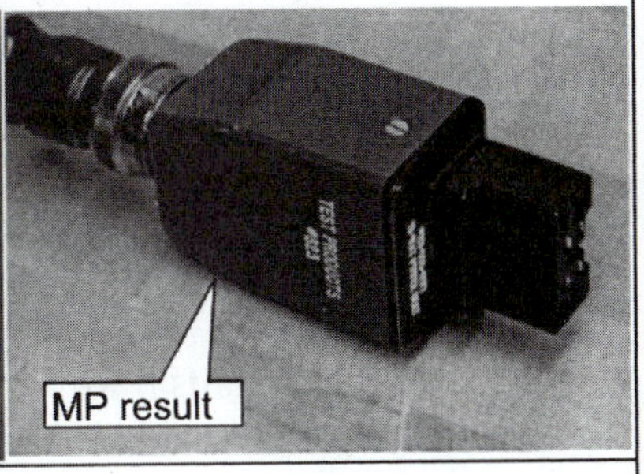

Original spare parts   MP result

Cause:
When the operator plugs the male connector into the female receptacle, only a small portion (3/8")of the connector is left exposed. Removing the plug from the receptacle is difficult and the wires are frequently pulled out the back of the connector

Countermeasure:
Provide a connector suitable for the application, where: 1) The connector body is easy to grasp and pulling on the wiring is not likely. 2) Provide strain relief on all connector bodies subject to handling and movement. 3) Quick change capability.

Effect of Improvement:
The operator will pull on the connector body to un-plug instead of pulling on the wires.

Engineering Dept.:	Date Received:	Accept:	Reject:	Other:	Signature:
Eng. Dept. Comments					

Date Returned to Maintenance:	G/L Signature:

FIGURE 7–5   FMEA used for hazard prevention and productivity improvement

7.6   ROOT CAUSE ANALYSIS

Root cause analysis is based on the premise that for every effect there exists a basic or root cause. In other words, events do not occur as a result of some random or chaotic process but are due to underlying causes or contributing factors. Root cause analysis is a systematic approach in identifying the basic or the root cause of a problem or an undesirable condition so that actions may be taken in order to eliminate that cause and prevent the occurrence of the undesirable event. In performing a root cause analysis, care should be taken so as not to confuse *symptoms* of a problem with the *cause* of a problem. Treating the symptoms neither removes the cause nor does it provide a real or lasting solution to the problem. In the case of a patient with a high fever, which may be due to an infection, reducing the fever may make the patient more comfortable but does not treat the infection. Similarly, replacing a leaking oil gasket is only treating a symptom of a more severe problem, which might have damaged the seal in the first place.

As the name and definition imply, root cause analysis has traditionally been applied as a problem-solving tool. After a problem or an undesirable event has occurred, root cause analysis is used to identify the most basic factor that may have caused or contributed to the situation. In addition to its traditionally reactive use, the astute reader as well as practitioners of root cause analysis quickly realize that it also can be applied in a proactive manner. As a logical step subsequent to FMEA, the root cause analysis process can provide significant insight into the cause or contributing factors that may lead to *potential* failure conditions.

Another important consideration in root cause analysis is the realization that although many causes can be identified initially, for all practical purposes there is only one or very few reasonable causes and very few feasible solutions available. For example, consider the power crisis that plagued California during spring 2001 and caused mandatory "rolling" blackouts. What would you consider as the root cause or causes of the immense electrical power shortage? The population of the state has grown steadily and significantly in the recent years; therefore, the added burden could be a reasonable root cause for the problem. How about the hot temperatures? That, too, could be a reasonably good contributing cause to the problem at hand. Other possible culprits for the root cause also could be listed. The proliferation of personal computers and other electronic devices made possible by current technology and a high standard of living certainly are taking their toll on the electrical power supply. All these factors are reasonable candidates for the root cause of the California power crisis.

The root cause, however, must be of such a nature that reasonable actions can be taken to eliminate the root cause and prevent the problem from occurring again. The root causes suggested previously all seem reasonable enough, but the solutions that could be developed from those causes are not realistic or practical. For example, we can neither control or avert the hot temperatures, nor can we determine the population of the state of California or how many personal gadgets residents of the state may possess. So what is the root cause? On June 27, 2001, the first new power plant in *13 years* came on line in California. It should be obvious to most casual observers that lack of proper foresight and planning, and failure in forecasting the trends in population growth and increased demand on the resources, are the most likely root causes of California's energy crisis.

Root cause analysis, applied in either a reactive situation to discover the causes of a problem at hand or in a proactive setting to predict the potential contributing causes to future adverse events, can significantly improve manufacturing and service system performance and reliability. Root cause analysis provides a means of systematically evaluating and understanding the details of the equipment and the system failure. By determining the root cause of failures, defects, accidents, and other undesirable events and then taking feasible corrective actions, the process of equipment "repair" can be turned into equipment "improvement." As a result, probability of future failure occurrences is drastically reduced or completely eliminated and productivity and quality are enhanced.

7.6.1   The Process of Root Cause Analysis

Figure 7–6 presents a schematic approach to root cause analysis. The first step in the process is problem identification and definition. In most manufacturing and business settings, finding a problem to solve is not difficult at all. There are plenty of them, and they will find you. And that is the challenge: selecting a problem that reasonably can be solved given the constraint of the available resources and time, and that also will provide an appropriate return on the efforts invested. In most cases, a problem is assigned to the analysis team. In other cases, some basic criteria should be used for problem selection.

Risk priority number (RPN) was discussed in the section about FMEA. Recall that RPN is calculated by multiplying the three rating values of occurrence, severity, and detection pertaining to a particular problem and is used to rank order or prioritize problems or potential failures. Therefore, a problem's RPN value is a reasonable criterion for problem selection. Events with high RPNs are strong candidates for selection.

Pareto analysis also can be used as selection process. Vilfredo Pareto (1848–1923), an Italian economist, made an interesting observation about the distribution of wealth in Milan that led to the vastly applicable and used principle of the 80-20 rule. He theorized that approximately 20% of the population possessed nearly 80% of the wealth, whereas the remaining majority owned only the remaining pittance.

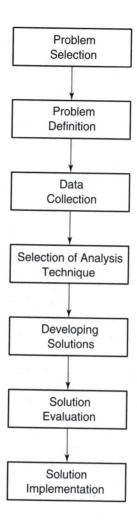

FIGURE 7–6 A schematic approach to root cause analysis

The rule has passed the test in many other situations over time. In this particular case, it can be generalized that in most cases, 80% of the downtime, for example, is caused by failure in 20% of the equipment; or, if we were to list all the problems or potential problems in a company, the top 20% of the problems would be responsible for nearly 80% of the losses. Therefore, Pareto analysis can be used to isolate the "important few" from the "trivial many." It is imperative to keep in mind that in the arena of continuous improvement, no problem or defect is trivial, but given the economic realities, it is prudent to solve the critical problems before the minor ones.

Problem definition Defining the problem is the most important step after problem selection. One cannot determine *why* (the root cause) something happened unless one can clearly define and state *what* (the problem) is. A clear understanding and statement of the problem is necessary before information about the circumstances of the incident can be collected. This is called forming a hypothesis—making an educated guess about what happened. The statement "the sensor failed," for example, may be a good educated guess as to why the conveyor did not stop or the robot did not properly position the part. Once data have been collected, the hypothesis is either confirmed or rejected.

Data collection Next to human resources, *pertinent* data is the most valuable asset to an organization. The emphasis is on pertinent, because if the collected data is not relevant, the information could be wasteful, expensive, and quite misleading. Data gathering is a time-consuming process, and therefore an expensive one. Too little information obviously is not good, but it is also possible to have too much information, so that we can no longer see the proverbial forest. Gather data for the specific problem at hand.

7.6.2 Techniques for Root Cause Analysis

The techniques for root cause analysis range from a simple unstructured process to more complex analytical methods. The most important objective to keep in mind is that root cause analysis is a problem-"solving," not a problem-"generating" process. The technique chosen should match the problem at hand and should be a means of clarifying the problem, not making matters more complicated. A tool should aid in the solution of the problem and not create a problem of its own. Whereas a complex problem with severe consequences may require a more formal and a detailed analytical technique, a straightforward problem could lend itself easily to a simpler and a less complicated process for resolution. With that in mind, we will proceed briefly to introduce and examine some of the commonly used root analysis techniques.

The common-sense approach This approach is less structured but just as valid as any other method of root cause analysis. It relies on the common sense, experience, and "knack" for doing their work of the individual team members. Their collective wisdom is pooled to analyze the data and the situation and offer a solution. Do not underestimate the intimate knowledge of operating personnel about the equipment and process condition and their innate ability to solve production and equipment-related problems given the opportunity to become actively involved in the process. Operator experience and intuition can play an important role in determining the root cause of a problem. We have all observed an experienced auto mechanic who can "hear" a specific sound or feel a particular vibration amid the roar of an engine and can identify exactly what the root cause of the problem is. Collective knowledge of the team and their ability to refer to their own past experiences or draw on the past history or experience of other plant personnel—what we may call in the new parlance of buzzwords "networking"—is an invaluable asset in pinpointing the root cause of various problems in an informal and unstructured fashion.

Perhaps the only word of caution regarding an informal process such as this would be to ensure that the team members keep an open mind and consider all possible alternative causes before making conclusions. Especially in the case of a reccurring problem, it may be that the team's "experience" has been accumulated through repeated treatment of the symptoms and not from curing the ailment.

Change analysis   The technique for root cause analysis that may be best suited for troubleshooting is based on the assumption that a change in the process, material, procedure, and so on must have occurred to cause the problem. If the eqipment or the process has been in a normal state of operation and is now experiencing some sort of difficulty, it is logical to assume that the abnormality or the trouble has been caused by some change. Therefore, the change analysis technique attempts to look for changes in various factors that might have an effect on the process or the equipment. Any change in the material, procedures, methods, tooling, human factors, and the environment can adversely affect the equipment or the process. Changes may act independently or interact with each other to produce problems in the system.

The purpose of the change analysis is to compare each factor before and after the difficulty has occurred and then attempt to determine what has changed. Sometimes changes are planned. Organizational changes, changes in material or processes due to engineering change orders or customer requirements, can be expected from time to time. In such cases, pinpointing the change as a possible source of the trouble does not necessarily create a challenge. Indeed, it would be prudent to be proactive and try to anticipate how the change could affect the system at the planning stage and then implement safeguards against any adverse effects that the changes may cause.

In other cases, the change is not planned and may occur inadvertently. Some serious detective work and ingenuity may be required to determine what has changed. These changes may be sudden, or they may be gradual and occur over a prolonged period of time.

In any case, change analysis requires a systematic approach in comparing the condition, setting, and state of any and all factors that could somehow influence the system before and after the failure and note any variances. Analyze these variances or differences and their possible causes to determine how these changes may have affected the system.

Figure 7–7 represents a sample form for change analysis. The first column is used to list all possible factors or variables that may influence the system. The second and the third columns are used to clearly describe the condition or the state of the factor before and after the failure occurred respectively.

Any marked or perceived difference or changes are noted in the fourth column. The last column is used for the analysis of the change and its possible effect on the process.

As stated earlier, a change by itself may not influence the process, but often a combination of two or more changes may bring about an undesirable change. In statistical design of experiments (DOE) for determining the effect of various variables on a given process, this phenomenon is referred to as the interaction between two (or more) variables. For example, changes in the feed rate and speed, if they occur separately, may not create a problem. In combination, however, their interaction may have an adverse effect on the system. Of course, the opposite also may be true: Changing one factor may necessitate a change in another factor to avert an undesirable outcome, or better yet, to improve the existing condition by increasing process yield.

The concept of design of experiments (DOE) was mentioned previously. Although an in-depth discussion of this subject is beyond the scope of this text, it might be of interest to the reader, to take a quick glimpse at the topic, in light of the subject at hand. Classical DOE is a statistical procedure introduced by Ronald Fisher in the early 1920s. The purpose of the technique is to isolate a series of variables that the experimenter suspects might be influential in a given process, and then systematically manipulate these variables and their settings to determine which variables and at which settings would optimize the process. One could reasonably argue that DOE is the most sophisticated level of change analysis, which attempts to anticipate the effect of changes on a process and take advantage of these changes in order to improve and optimize a system.

Events and causal factors analysis   The chaos theory has given rise to the question that ponders, if a butterfly flapped its wings in the jungles of South America, how would that affect the weather halfway around the globe? The question obviously makes a point that events do not simply happen—there is always a causal relationship, a cause and a resulting effect. Events refer to incidents, system failures, and other similar occurrences. Causal factors are those contributing variables or features that have caused the event to occur or have contributed to its

Process Variables	Conditions Prior to Failure	Conditions After Failure	Variance	Analysis

Equipment:
  Feed
  Speed
  Alignment
  Load
  Etc.				

Material:
  Specs
  Type
  Etc.				

Operator:
  Training
  Attitude
  Experience
  Etc.				

Environment:
  Temperature
  Humidity
  Seasons
  Etc.				

Other Possible Variables:				

FIGURE 7–7   A sample form for performing change analysis

occurrence. Event and causal factors analysis is a systematic approach used to examine these events, their sequence, the conditions under which the events occurred, and any factors that may have resulted in or caused that event.

An event and causal factors diagram can help the investigator establish, visualize, and understand relationships among various events and their contributing factors. It also can facilitate the discovery of missing links and information. Two methods are used to develop a diagram for this purpose. We will examine a standard event and causal factors diagram in this section. The other method, using an Ishikawa diagram, is covered in the next section of the chapter.

Figure 7–8a represents an event and casual factors diagram. Rectangles are used to represent events, and conditions or factors contributing to these events are depicted with ellipses. Events or conditions that are based on actual and valid observations and are certain to have occurred are drawn with solid lines. Some analysts use dashed lines to represent events or conditions that are not based on factual data and are assumed to have occurred. The inclusion of unsubstantiated and assumed data can lead to erroneous conclusions, however, and is discouraged. Solid

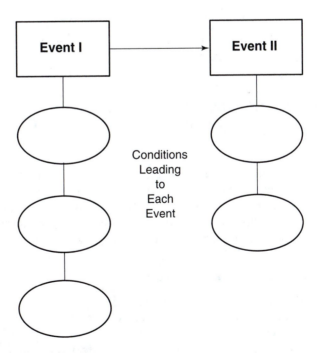

FIGURE 7–8a   An event and causal factors diagram

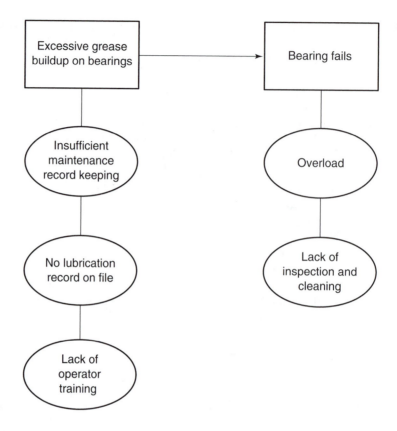

FIGURE 7–8b   A simple example of an event and causal factors
diagram

lines are used to connect factors to their resulting events. The diagram also
presents the chronology or sequence of events. The flow of events is from
left to right; therefore, older events are placed on the left and more recent
ones follow to the right. Simple and complex examples of event and causal
diagrams are presented in Figures 7–8b and Figure 7–8c respectively.

7.6.3   Ishikawa Diagram

Ishikawa diagrams are named for their developer, Kaoru Ishikawa, a
Japanese pioneer in the field of quality assurance. They can be used to
represent the relationship between an event and its causal factors and
hence are known as cause-and-effect diagrams. Due to its distinctive
shape, the diagram is also referred to as the fishbone diagram.

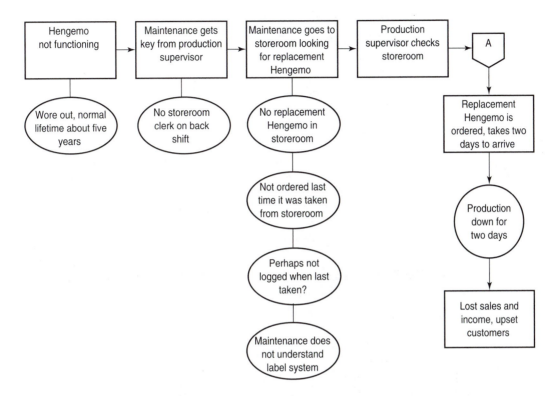

FIGURE 7–8c　A detailed example of an event and causal factors diagram

Figure 7–9a represents a generic cause-and-effect diagram. The "head" of the fish represents the event or the effect (the problem). Each branch can be used to present a major factor or cause leading to the event. Each major branch can have smaller divisions or branches representing smaller contributing factors that are subordinate to the major factor. Figure 7–9b represents a fishbone diagram for a hypothetical system failure. Various factors such as equipment, personnel, material, and so on may have caused the failure to occur. Major branches represent major categories or classifications of causes for the failure. The causes generally are referred to as the six "Ms": Material, Machine (equipment in general), Measurement (any measurement errors or problems caused by measurement instruments), Methods (work or process methods that may be too complicated or unclear), Manpower (personnel), and Mother Nature (environmental factors). Similarly, smaller branches under each major category represent those variables that might have contributed as a cause of failure to that causal factor. For example, subordinate to the major factor "personnel," possible

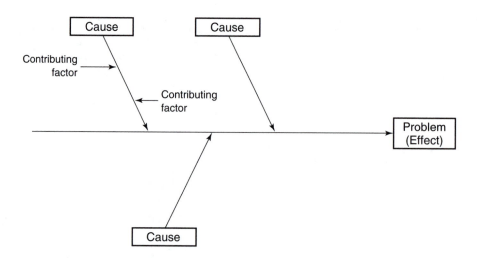

FIGURE 7–9a A generic cause-and-effect diagram, also known as an Ishikawa diagram

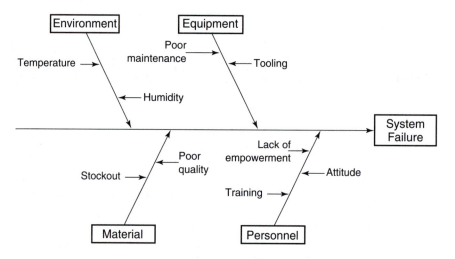

FIGURE 7–9b A cause-and-effect diagram for a hypothetical system failure

contributing factors may include insufficient or inadequate training, lack of empowerment, less-than-desirable attitude, and so on. An Ishikawa diagram is an excellent graphical tool for brainstorming and identifying all the possible causes of the problem for further investigation.

7.7 FAULT TREE ANALYSIS

An excellent tool to visualize and understand more complex relationships, fault tree analysis is a graphical display that aids in constructing a clear and logical association between an event and its contributing factors. Although it is similar to the cause-and-effect diagram, fault tree analysis adds the dimension of "conditional requirements" among the contributing factors and shows the requisite conditions under which an event might take place. Fault tree analysis is helpful in understanding that the presence of a given condition by itself may not be a sufficient cause for an event to occur. Conversely, it shows how various factors in the presence of other factors may exacerbate and contribute to a system failure.

Although an array of symbols are available for constructing a fault tree, most systems can be defined easily and clearly with the use of three basic symbols: rectangles, "OR" gates, and "AND" gates.

Rectangles are used to define any events, factors, or causes. Two or more lines from each "OR" gate lead to two or more factors. "OR" gates are used to indicate that *any* one of the factors or causes that follow is a sufficient cause by itself to produce the effect. Therefore, if any one or more of these factors are present, the event is likely to occur. For example, an automobile will fail to start if the starter OR the battery fails. Figure 7–10a shows an "OR" gate.

"AND" gates, as shown in Figure 7–10b, are used to indicate that *all* the causes or factors that follow must be present in order for the event to happen. As with the "OR" gates, two or more lines from the "AND" gates lead to the requisite conditions. Contrary to the "OR" gate, however, the presence of one or more factors does not create a sufficient cause—*all factors must be present.* For instance, if either the emergency

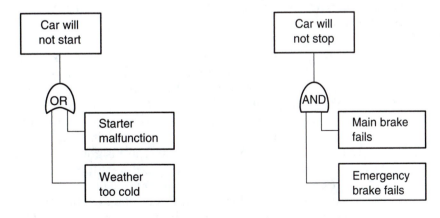

FIGURE 7–10a "OR" gate   **FIGURE 7–10b** "AND" gate

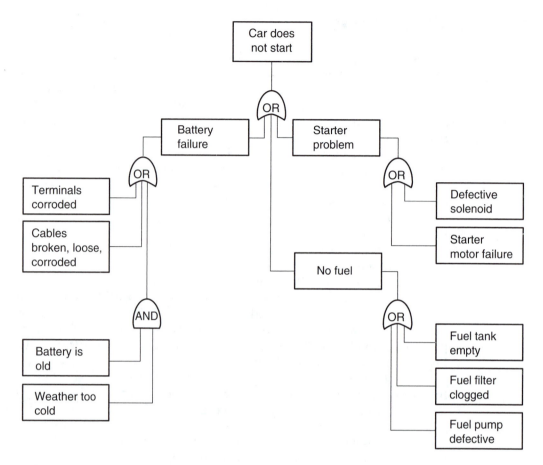

FIGURE 7–11   A fault tree analysis

(Source: Example development and setup courtesy of James R. Kimmey, professor emeritus, Elgin Community College.)

brake or the main brake functions, we can stop a moving car. Therefore, the failure in stopping the car requires that the main brake AND the emergency brake both fail. A detailed example of fault tree analysis utilizing a series of "AND" and "OR" gates is shown in Figure 7–11.

The construction and analysis of the fault tree shows all the causes, their relationships, and the necessary conditions that can result in a failure. The analysis can provide a more clear insight into the process and how to develop safeguards and redundancies to reduce the probability of such failures. Fault trees can be used as a troubleshooting tool to gain insight into an existing problem. During the design stages, fault tree analysis can bring to light any process or product shortcomings and help with the design review process in order to eliminate any potential for future failures.

More elaborate fault tree structures can make use of historical and reliabilities studies to determine the statistical probability associated with each condition or factor. These numerical values then can be used to predict the likelihood or the probability that each effect will occur and the overall probability of system failure. These quantitative data can help investigators determine, with a known level of certainty or confidence, the probability of each failure, the cost associated with each failure, and the cost of preventive measures such as built-in redundancies, so that they can make educated and informed decisions about how to solve a specific problem.

7.8   IMPLEMENTING A SOLUTION

In summary, regardless of the methods employed in the process of root cause analysis, the primary objective is to find a solution to the potential problem or the problem at hand. As stated previously, the most important input into the root cause analysis is the problem definition. The root cause analysis, whether informal or formal and structured, attempts to discover the conditions or factors that contribute or cause a failure or a potential problem. The purpose, or the most important output of root cause analysis, is to devise a solution or solutions that will remove the hazard, resolve the problem, and ideally prevent its future occurrence. For if we fail to arrive at a solution or if we are not diligent in implementing such a solution, we will have engaged in an exercise in futility.

7.8.1   Alternative Solutions

Often we may arrive at several alternative solutions. Some are more realistic and feasible than others. Some are long-term solutions, whereas others may be temporary and serve as a quick fix. No solution may be perfect and none should be rejected offhand. Every solution must be evaluated based on the data available.

Although long-term solutions are obviously preferable, they may require a longer period of time to implement. If time and resources are not available, a short-term solution or a temporary fix might be necessary to deal with the problem at hand. The problem of the energy crisis in California that was discussed earlier serves a useful purpose here. Nobody can argue with conservation and more efficient use of energy as definite solutions. But is that a long-term solution? As the population grows and the demand increases, in the long run the problem will continue. A long-term solution might be the development of additional and alternate sources of energy. The latter is probably a longer-term solution, but what can be done until these new facilities or alternatives are put in place?

As with the statement of the problem, each solution should be clearly defined in terms of its scope, goals, objectives, attributes, and the various metrics or measures that are used for its evaluation. These measurement criteria, or assessments, are essential to determine the appropriateness and effectiveness of the proposed solution.

Alternative solutions may be evaluated based on implementation costs, feasibility, return on investment, ease and difficulty in implementation, constraints and limitations, and of course, the probability that the given solution will be successful in removing the hazard or eliminating its future occurrence.

7.8.2 Evaluating Alternative Solutions

Various methods can be used to evaluate alternative solutions. Laboratory tests and use of prototypes allow experimentation with physical and tangible models. If the solution requires design modifications to products or processes, physical tests provide the most realistic environment for evaluation. Classic examples include crash testing of passenger vehicles to test and evaluate various safety features such as seatbelts, bumpers, and airbags. Accident and crash recreations can be thought of as the reverse of the process of evaluation used to determine the root causes or factors contributing to the accident.

Sophisticated and advanced software programs have made the use of computer simulation and modeling for evaluating alternative solution a viable technique. A variety of alternatives and scenarios can be simulated using animation and real-life characteristics. A major advantage of computer simulation is the nondestructive nature of the testing. Another advantage of computer simulation is its ability to accelerate the time factor of the process. When the effect of time on the process must be evaluated and the evaluation process requires a significant passage of time, computer simulation can speed up the process and simulate the passage of hours, days, and years in matter of just a few minutes or hours.

Mathematical and statistical models not only enable us to clearly and quantitatively visualize different production scenarios, they also serve as a powerful tool in evaluating various alternatives. The novice user quickly develops a keen sense of appreciation for these tools. All engineering solutions must lend themselves to mathematical evaluation and confirmation. Regression and trend analysis, statistical design of experiments and factorial analysis, and a host of other statistical and mathematical procedures are the proven workhorses of testing and evaluating alternative solutions.

We would be remiss at this point if we fail to mention an obvious yet extremely important aspect of *any* solution. The operating personnel, the team, all the employees, and the customers who are expected to be involved in the implementation of the solution or are affected by the solution must be an integral part of the solution. If these individuals are included in the process, they will also be part of the solution and will take ownership of the solution. Otherwise, they will be a part of the problem and *your solution* will fail.

A CHAIN IS ONLY AS STRONG AS ITS WEAKEST LINK

Subaru-Isuzu Automotive, Inc. (SIA) was briefly introduced in the previous case study in this chapter. SIA's outstanding TPM program has led to significant gains in quality improvement and overall equipment efficiency increases. SIA's investments in educational programs for its employees and management commitment, along with the employees' involvement, participation, and teamwork, are the primary reasons for such great achievements.

Examples of continuous quality and productivity improvement activities are abundant. In this case study, three separate and distinct examples of root cause analyses are presented in Figures 7–12 through 7–14.

The case documentation for each example is self-explanatory and requires no additional commentary. However, it may be helpful to restate the steps in root cause analysis and solution recommendation processes briefly here.

The top section of the form documents the time, date, location, and other pertinent information. This information can be used as part of the equipment history and to determine cost, OEE, and so on.

As required in root cause analysis, a brief statement clearly and concisely describes the problem or the phenomenon. Asking repeated "whys" to get to the root cause of the problem is an effective root cause analysis method. Asking five or more "whys," each progressively more probing, tends to help get past the symptoms and find the root cause so that a final solution can be reached. In the second analysis presented here, the root cause was discovered after only four questions. In all three of these root cause analyses, photographs enhance the visualization and the resolution of the problem.

The implementation of the permanent solution often may require additional time. In such cases, a temporary solution may be suggested to continue production activities in a safe manner until the final solution can be implemented. In these three cases, temporary solutions preceded the permanent solutions.

Additional information on the sample forms clarifies the need for any specific further action, the date for such actions, and the team or the individual who is to perform such actions. Follow-up actions also are documented. Figure 7–15 presents a schematic approach to the root cause analysis used in these cases. Table 7–4 outlines the problem tracking analysis guide and Table 7–5 lists the questions used in this process.

(continued on next page)

Date of Occurrence	2/4/2000	Report Over in Min. Type (15) or (30)	30 min	Tracking #	trim-suv 2-4-00a
Shift	first	Line Name	Engine sub-assembly		Ranking A
Stop Time	6:15am	Equipment Name	Anchor engine sub conveyor		
Start Time	7:00am	Written by:			Were Parts available Y=yes
Total D/T	45 min	Repaired by:	entire first & third shift		
Units Lost	11	Repair in Man hours (M x H = MH)	7 x 1 = 7 manhours		

Problem / Phenomenon
Condition of Occurrence

Empty pallets would not return. Found chain on north drive was bunched up inside the tubular frame.

5-Why
Actions taken to find root cause

1 The drive stopped because the chain fell apart and bunched up on the return rail.

2 The chain came apart because the master link broke.

3 The master link broke because the side plates of the chain separated.

4 The side links seperated because the cotter pin was sheared away.

5 The cotter pin sheared because lateral forces exceeded its design strength.

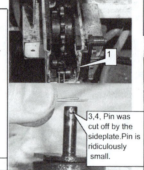

2 The remnants of the pin inside hole confirm the pin did not fall out.

1

5 Twisting action likely due to roller offset.

3,4, Pin was cut off by the sideplate.Pin is ridiculously small.

Temporary Countermeasure
Immediate action / may re-occur

Dis-assemble conveyor enough to unjam the chain,which was wadded up inside the 3" X 6" tube. Replace master link and resume operation. The fact that the chain returns inside a tube impossed a repair time penalty (when the chain breaks), of approximately 20 minutes in this case.

Permanent Countermeasure
Final Action / Never to occur again

It appears the chain is really the biggest problem and the master link is the weakest link. A chain with stronger master links is not compatible with the compact design, so over torque "shock relay" was added.

Further actions needed:	Yes	No	Associate in charge	Target Date	Date Completed
1.) Add to Maintenance P.M. Standards		x			
2.) Add to Production TPM Standards		x			
3.) Make Troubleshooting Step Checklist		x			
4.) Apply Countermeasures to like Equipment	x			3/30/2000	3/18/2000
5.) Request assistance from Mfg. Engineering		x			
	Signature:				

FIGURE 7–12   Root cause analysis for conveyor chain malfunction
(Source: Courtesy of Subaru-Isuzu Automotive, Inc.)

Date of Occurrence	5/2/2000	15 / 30 min. Report	**30**		Tracking # →	00-tt-023

Shift	1	Line Name	Shower test	**Ranking**

Stop Time	2:00pm	Equipment Name	TF-202i	☐ AA ☑ A

Start Time	9:30pm	Written by:		☐ B ☐ C

Total D/T	7.5hours	Repaired by:	primarily 2nd shift	**Parts Available**

Units Lost	0	Repair in Man hours (M x H = MH)	32 hours	☑ Yes ☐ No

Problem / Phenomenon

The conveyor stopped moving, the shear pin had broke and a slat jammed in the drive area.

5-Why Analysis

1 The conveyor stopped moving because the shear pin broke.

2 The shear pin broke because the wheels jumped the track near the drive motor.

3 The wheels jumped the track because a slat plate jammed on the top of the track.

4 The slat plate jammed because of too much slack in the conveyor. The bolts were broke off on the slat plate attachment bracket.

5

Wheels forced under the track by slat plate.

The wheels belong on top of the track.

Temporary Countermeasure: Immediate action / may re-occur

Unjam and reset the conveyor on track, replace broken shear pin and bent slat plate. Adjust take-up to proper tension.

Permanent Countermeasure: Final Action / Never to occur again	Associate in charge	Completion Target Date	Follow Up **CHECK** by: <u>Manager - G/L</u>		
			1 mo.	2 mo.	3 mo.
Add 6 month inspection/adjustment to pm sheets of this and similar equipment. Replace the shear pin with one to break before so much damage can occur. TPM activity to remove mounting bolts, apply medium threadlocker and re-attach the plates.	Shear pin selection	5/27/2000			

Further actions needed:	Yes	No				
1.) Add to Maintenance P.M. Standards	☑	☐		5/13/2000	done	ok	ok
2.) Add to Production TPM Standards	☑	☐		done	ok	ok	ok
3.) Make Troubleshooting Step Checklist	☐	☑					
4.) Apply Countermeasures to like Equipment	☑	☐		5/13/2000	done	ok	ok
5.) Request assistance from Mfg. Engineering	☐	☑					

Signature _____

FIGURE 7–13   Root cause analysis for broken shear pin

(Source: Courtesy of Subaru-Isuzu Automotive, Inc.)   *(continued on next page)*

Date of Occurrence	3/23/2001	15 / 30 min. Report	**30 minutes**	Tracking # ➝	01-tt-101

Shift	1	Line Name	TF-803	**Ranking**	
Stop Time	6:52am	Equipment Name	TP-852	☑ AA ☐ A	
Start Time	7:45 AM	Written by:	Jim H	☐ B ☐ C	
Total D/T	54 min	Repaired by:	3rd & 1st shift	**Parts Available**	
Units Lost	27	Repair in Man hours (M x H = MH)	4 X 60 = 240	☑ Yes ☐ No	

Problem / Phenomenon

TF-803 quit running, all the lights in the panel were off.

5-Why Analysis

1 All the lights in the panel were off because the fuses in the overhead buss disconnect were blown.

2 The fuses blew because the conveyor drive brack rectifier blew out (see photo).

3 The rectifier blew out because a metal object shorted terminal 4 of the rectifier to ground.

4 Metal objects shorted out terminal 4 because the motor junction box was full of nuts, bolts, metal tear tabs, etc...

5 The juction box was full of auto parts because any parts dropped on the conveyor falls off at the drive end. One screw was missing from the motor box & the cover turned sideways, allowing parts into the wiring box.

Short from screw teminal to ground.

j- Example of cover half on. Parts fell in, shorting open screw terminal on rectifier.

Yes / No

Temporary Countermeasure: Immediate action / may re-occur

Get snorkel lift and replace fuse. Disconnect brake rectifier and install cover.

Permanent Countermeasure: Final Action / Never to occur again	**Associate in charge**	**Completion** Target Date	Follow Up **CHECK** by: Manager - G/L		
			1 mo.	2 mo.	3 mo.
Replace the brake rectifier, Make a guard to deflect the mass of parts falling of conveyor away from drive.		24-Mar			

Further actions needed:	Yes	No			
1.) Add to Maintenance P.M. Standards	☐	☑			
2.) Add to Production TPM Standards	☐	☑			
3.) Make Troubleshooting Step Checklist	☐	☑			
4.) Apply Countermeasures to like Equipment	☐	☑			
5.) Request assistance from Mfg. Engineering	☐	☑			

Signature *Jim H*

Mgr's Signature _____ Date _____

G/L's Signature _____ Date _____

14-CTP4-003-1.0

Issue Date 5/1/2000

FIGURE 7–14   Root cause analysis for an unplanned downtime
(Source: Courtesy of Subaru-Isuzu Automotive, Inc.)

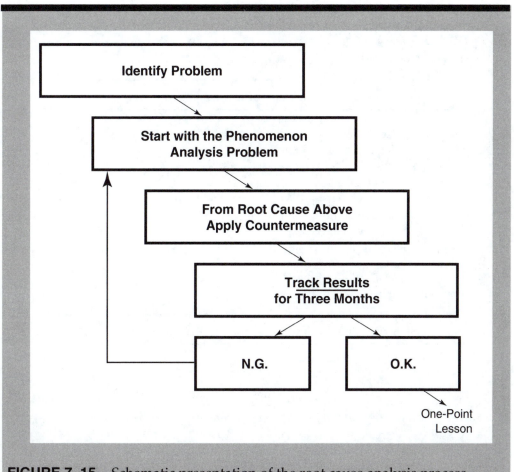

FIGURE 7–15   Schematic presentation of the root cause analysis process
(Source: Courtesy of Subaru-Isuzu Automotive, Inc.)

(continued on next page)

TABLE 7–4
Problem Tracking Analysis Steps

Step One: Clarify the phenomenon

Eliminate preconceived notions.
 Clearly specify items to be observed to avoid errors based on supposition and rigid thinking.

Carefully observe and analyze the *facts* on the shop floor.
 Trace the problem back to the smallest possible units.
 Personally observe its physical elements at the site.

Sort the phenomena thoroughly using five Ws and one H.
 Classify what you have observed.

Compare the good products with the defective products to pinpoint all significant differences.
 Investigate deviations; clarify distinction between what is normal and abnormal in observation.

Step Two: Conduct a physical analysis—Think visually

Identify operating principles.
 Review machine diagrams and manuals to understand the equipment's basic operation.

Identify operating standards.
 Learn function and mechanism of equipment/devices using simple sketches.

Identify interacting elements.
 Draw diagrams to identify what relationships define the phenomenon.

Quantify the physical changes involved.
 Identify appropriate physical quantities and changes in those quantities.

Source: Courtesy of Subaru-Isuzu Automotive Inc.

TABLE 7–5

Five Ws and One H

Who:	Any variation among people involved in the operation? (Any morning/day/night shift difference?) (Any difference among new operators/floaters/temps?)
What:	Any variation due to production materials? (Any material differences?) (Any variation due to part dimensions?)
Where:	Any variation due to equipment, fixtures, components? (In what process/machine elements does the problem occur?) (Any differences among different equipment/machines?) (Any variation associated with different jigs/fixtures?)
When:	Any variation due to time or period? (Does problem occur at the start of work? In the middle?) (Any time differences associated with the problem?) (During which operations is the problem likely to occur?) (Is the problem likely to occur after setup changes?)
Which:	Are there any characteristic trends over time? (Do problems increase or decrease?) (Any changes before, after, or simultaneously?)
How:	Any variation in circumstances of occurrence? (Does the problem occur frequently or only rarely?) (Does it appear abruptly or gradually?) (Does the problem appear continuously or at random?) (Does it appear at regular or irregular intervals?)

Source: Courtesy of Subaru-Isuzu Automotive Inc.

Subaru-Isuzu's relentless commitment to total productive maintenance is based on a solid foundation of strong management support and
employee empowerment and pride. As has been stated repeatedly, TPM
is more than a set of rules listing do's and don'ts. It is a culture. Once the
culture has been established and the management shows its genuine
commitment and direct involvement, followed by the education and empowerment of the employee, TPM becomes an inseparable and integral
part of every employee's way of doing work and every day's activities—
it becomes a way of life.

To illustrate managements' direct involvement and commitment,
the author recalls his first visit to Subaru-Isuzu. As the author noted the
immaculate condition of the entire facility (TPM's rule # 1: Clean, Clean,
Clean!), the senior vice president for manufacturing commented that the
conditions had been drastically different in the pre-TPM days. Of course,
looking at the current spotless facility, it would have been hard for anyone to imagine that there had ever been a speck of dirt anywhere in the
plant, despite the repeated assurances of that fact. But the important lesson here is that once the plant had made a commitment to TPM, the involvement and commitment started with the top management. The top
brass, in their coveralls, worked alongside the shop floor employees,
cleaning and polishing brass. This is TPM at its finest hour.

TPM is an ongoing and never-ending process. It is integral in every
day's activities. An employee does not stop doing a task in order to do
TPM—that worker includes TPM as part of every action and every task.
The result is continuous problem and failure prevention and avoidance.
Figures 7–16 through 7–19 illustrate a brief yet clear story, the story of
constant vigilance to detect and solve potential problems *before* they occur. Tear and wear are normal progressions for any object, but the ability
to detect and avert potential failure modes is the success story of TPM.
Figures 7–16 through 7–19 tell the story of every employee who takes a
stake in the ownership of the process and facilities and becomes a member of the TPM team.

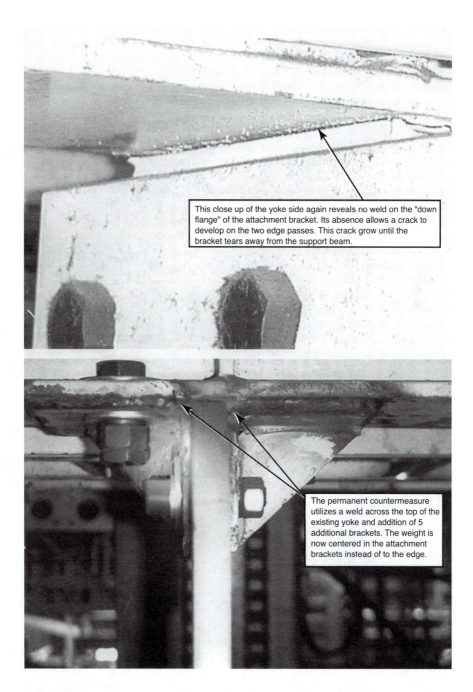

This close up of the yoke side again reveals no weld on the "down flange" of the attachment bracket. Its absence allows a crack to develop on the two edge passes. This crack grow until the bracket tears away from the support beam.

The permanent countermeasure utilizes a weld across the top of the existing yoke and addition of 5 additional brackets. The weight is now centered in the attachment brackets instead of to the edge.

FIGURE 7–16   A serious catastrophe is averted by inspection and PM
(Source: Courtesy of Subaru-Isuzu Automotive, Inc.)

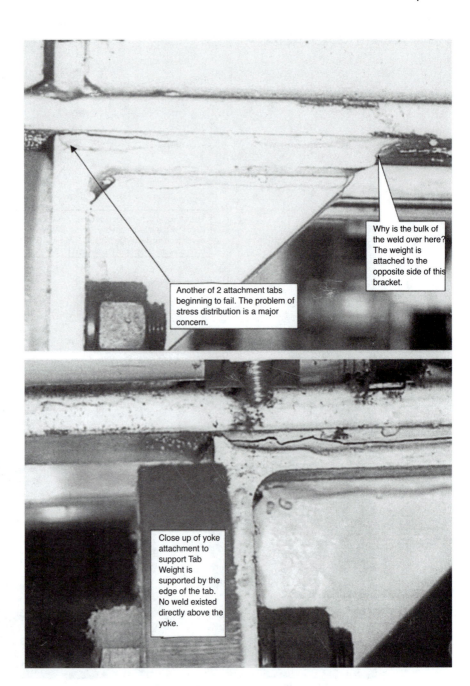

FIGURE 7–17   Another potential accident is discovered as a result
                    of PM inspection
(Source: Courtesy of Subaru-Isuzu Automotive, Inc.)

Support structure for TD-806. "FABA" motrain is supported by 11 yokes, no span exceeding 5 feet. 4 in a row broke, all within 15 feet of one another.

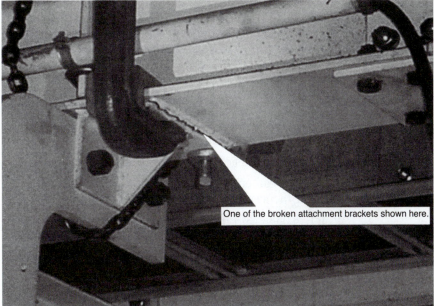

One of the broken attachment brackets shown here.

FIGURE 7–18   Regular PM inspection aids in disaster avoidance
(Source: Courtesy of Subaru-Isuzu Automotive, Inc.)

Aisle traversing motrain X30 Y14.

FIGURE 7–19   Another potential problem that could result in
                  downtime is revealed as result of regular PM
(Source: Courtesy of Subaru-Isuzu Automotive, Inc.)

QUESTIONS

1. Define benchmarking.

2. What is the basic blueprint set forth by AT&T for benchmarking?

3. How does benchmarking help a company improve their methods
   and practices?

4. It has been said that those who practice benchmarking are doomed
   always to be the second best. How would you agree or disagree
   with this statement?

5. What are the five pillars of TPM?

6. What are the eight pillars of maintenance?

7. Clearly define and explain the concept of FMEA. What role can
   FMEA potentially play in TPM?

8. Based on a full understanding of the concept of FMEA, how would
   you rename FMEA?

9. What are some of the potential failure modes affecting your calculator? What effects could these failures have?

10. Use Tables 7–1, 7–2, and 7–3 to calculate the RPN for the following situations:
    a. severity: high; occurrence: moderate (5 per 1000); detection: very high
    b. severity: very minor; occurrence: very high; detection: very remote
    c. severity: low; occurrence: remote; detection: moderate

11. Rank order (high to low) the situations from Question 10 based on their priority.

12. Define root cause analysis. What is the significance of root cause analysis in maintenance?

13. Perform a root cause analysis on a task for which you may have had less than a desirable outcome.

14. What role does Pareto analysis play in root cause analysis?

15. Explain the "change analysis" approach to root cause analysis.

16. What is design of experiments (DOE)? Do you see an overall relationship between DOE and maintenance functions? (*Hint*: Think about root cause analysis.)

17. What is an "events and causal factors analysis" and what does a butterfly flapping its wings have to do with anything, anyway?

18. How would an Ishikawa diagram help with a root cause analysis?

19. Explain an "OR" gate and an "AND" gate. How do they differ?

20. What is a fault tree?

21. Select a simple problem that you may have encountered in the past day or two, for example, your PC did not start, your electric razor did not work, your curling iron did not heat, or so on. Set up a fault tree for the problem.

22. Explain what is meant in this chapter by the statement that if the workers are not part of the solution, they will be part of the problem.

23. Explain some of the methods of evaluating alternative solutions.

24. Auto manufacturers, insurance companies, and consumer agencies, among others, spend millions of dollars annually to test various products by destroying them, that is, by crash testing them. Explain what they intend to accomplish with these tests.

8

FACILITY MAINTENANCE PROJECTS PLANNING AND CONTROL

Overview

Objectives

At the completion of the chapter, students should be able to

- Distinguish projects from repetitive tasks.
- Understand the key elements in successful project management.
- Identify specific project manager responsibilities.
- Generate a high-level project plan and a detailed work plan.
- Prepare a rough estimate of project costs early in the project.
- Draw a network diagram.
- Perform a crash analysis.

8.1 FACILITY MAINTENANCE PROJECTS

Maintenance projects are often characterized as repetitive and daily routine maintenance tasks, including preventive and normal everyday maintenance. Many maintenance projects are long-term by nature, however, and managing these projects is best accomplished using project management techniques. The preceding chapters focused on maintenance management best practices to facilitate successful facility maintenance planning. This chapter describes how to organize long-term maintenance work into separate tasks and monitor the progress of the maintenance work performed. We then describe specific project management skills and techniques, including the use of PERT/CPM.

8.1.1 Definition of a Maintenance Project

Projects are generally one-of-a-kind groups of activities that perform a specific function. Maintenance projects can be defined by the following characteristics:

- They require a group of tasks to be performed concurrently and/or consecutively.
- They can be defined by the dollar value of the activities performed.
- They are defined by the level of effort required by multiple departments in an organization.
- They require more planning and design effort than normal maintenance and preventive maintenance activities.
- Resources are allocated among tasks to accomplish the project within time and budget constraints.
- They could be delayed to a future date.

Let's look at an example of a maintenance activity that can be characterized as a project. A construction company was contracted to install a 44-mile, 12-inch-diameter pipeline to carry heated water and soda ash solution from a mining site to a processing plant. This "project" required many activities to be performed concurrently, such as using explosives to clear a path while contracting the purchase of heavy equipment for lowering the pipeline later.

The project was presented to the customer with a bid price to perform the activities. The construction company worked with many types of workers to perform the project, from explosion experts and fencing crews to construction experts and contract personnel. The 44-mile stretch between the mining site and the processing plant included mountainous terrain at slopes of up to 40 degrees and flat terrain used for farmland. The farmland contained many different types of electrical lines and pipes.

Naturally, this added complexity required much advance planning to ensure the project's success. This project had so many specialized activities that the construction company had to allocate the tasks carefully among all its specialized resources to complete the project on time and under budget. Lastly, this project was an improvement from the previous method of transporting heated water and soda ash solution between the two sites. Consequently, the project could have been delayed to a future date, as it was an optional activity pursued to reduce the long-term transportation costs for the mining company.

8.1.2 Successful Project Management

Successful completion of facility maintenance projects requires unique managerial skills and management of personnel from various functional departments, such as engineering, laborers, inspectors, procurement specialists, and schedulers. Selecting the best personnel for the project team can be tricky, as the project manager wants to recruit the most talented employees, but departments are generally reluctant to give up their best employees to the project team. A key element in project management is to facilitate upper-level commitment by defining the work by the results or benefits the company will receive. Three key areas need to be evaluated:

1. What are the project objectives?

2. What is the process to be used to complete the project?

3. What will the result of the project be?

Remember, we are talking about successful project management. So the first step is to identify what the goals are for this project. Naturally, to be successful, the project manager must identify what the customer is expecting as a result of this project. Are they expecting a piece of equipment to manufacture perfect quality parts at a speed twice as fast as the previous machine? Or are they expecting a machine that can perform completely new functions? The project can be successful only if we identify the expectations of the end user (customer) and manage these expectations throughout the project. Additionally, we must define for ourselves (the project team) and the customer why we should complete this project. What is the goal we hope to achieve at the end of the project? And if we didn't complete the project, what would be the consequences?

The process by which we complete the project also must be evaluated. We will describe this process in more detail later in this chapter, but initially we must identify the work tasks that need to be completed. Just knowing that the farmland must be cleared in the earlier example is not enough. We must determine when each of the activities should be completed and assign workers to each of the activities based on their skill level, cost, and availability.

The results also must be clearly defined. This means determining the specific benefits of the project and ways to measure the value of these benefits.

In summary, answer the following questions to help guarantee the success of a project:

Objectives:	What are the expectations?
	Why should we complete this project?
Process:	What needs to be done?
	When do we need to do it?
	Who should do it?
Results:	What are the benefits of the project?

8.2   THE ROLE OF THE PROJECT MANAGER

Managing projects requires a unique set of skills. The success of a project depends on the ability of the project manager to coordinate the people involved in the project, as well as his or her knowledge of the technical project management side of the project. Project teams are generally made up of a diverse group of individuals with unique skills and unique personality traits. This team of people must be managed to ensure the success of the project. What are the tasks required of the project manager? Are there specific talents and abilities required to guide a project team to success?

8.2.1   The Project Manager's Responsibilities

The project manager has responsibility for setting up the project schedule, controlling the steps in the project, and managing the team. Let's examine the project management process and determine the keys to controlling the project and managing the team.

Setting up the project schedule   The first job of the project manager is generally to set up the project plan. The following steps are key to the success of any project:

1. Clearly define the project scope, specifications, and deliverables.
2. Identify all activities for the project.
3. Identify any precedence relationships for tasks.
4. Identify resources available and assign personnel to the project.
5. Develop time estimates for each activity.
6. Develop the project schedule using Gantt charts.
7. Identify the longest path of the project.
8. Identify any slack for the project.
9. Manage the project.

Before beginning any project, as we discussed earlier in the chapter, it is important to clarify the project scope. Identify the project goal so that the customer and the project team agree on what will be achieved at the end of the project. The project scope includes identifying how long the project will take, what will be delivered to the customer at the end of the project, and the extent of customer involvement necessary to complete the steps in the process. Even the most carefully planned projects have customer revisions requested before the project is completed. There are many justifiable reasons for making changes, but as the project manager, managing the change process is very important. The project manager must ensure that all deliverables, budgets, schedule, and cost impacts are discussed with the customer *before* agreeing to incorporate the changes into the project. Specifications include any performance requirements, such as that a machine should be able to produce 100 parts per minute after an overhaul. The deliverables specify exactly what the customer will receive at the end of the project.

The project manager, with the help of key team members, identifies all of the specific activities needed to complete the project. Each activity name should be succinct and clearly stated. Start with the major phases of the project and then identify several subprojects within each major phase. This strategy is commonly referred to as the work breakdown structure. For example, in Figure 8–1, we identify four major phases or stages after the project is initiated.

These phases or stages are commonly referred to as the assessment stage, the formative/refining stage, the implementation stage, and the maintenance stage. Then we identify all the specific tasks that go under each stage. For example, the assessment stage begins with Survey Questionnaire Development and ends with Focus Groups. For these tasks, we also must determine which tasks must be completed before another task can start. This is called identifying a precedence relationship. For example, if we were working on a project that consisted of building a new factory, obviously clearing the land must precede digging the foundation, which must precede framing the building.

For each task, identify who should be responsible, that is, who is the resource for that task. As a project manager, you might be able to select the personnel on the project. However, sometimes you must do the best you can with the resources assigned to you.

Successful project management depends on proper time estimates. We discuss estimating time for project steps later in the chapter.

Using the tasks and the project management techniques discussed later in the chapter, the project manager must develop the project schedule, frequently using a Gantt chart (see Figure 8–1). Gantt charts display project activities using time scale representation. For each task identified in the previous section, the start and completion dates are identified and displayed on the horizontal bar chart at the right in Figure 8–1. To draw a Gantt chart, the tasks and their precedence relationships must be identified. A bar is drawn showing the time expected to complete each task. Revising Gantt charts can be cumbersome, as actual time durations fluctuate during the project. For this reason, most project planning is accomplished using project management software packages such as QSB, Microsoft Project, or other software tools.

Once all the activities are defined, the precedence relationships are identified, the activity times are estimated, and the resources are assigned, then you can determine the longest path of the project. Add up each path and the longest path becomes the critical path, in that its completion determines the total length of the project. This path will determine whether the project can be completed on time. Any slippages along the critical path might result in the project being late, depending on how much slack there is between the total length of the critical path and the time allotted to complete the project. The project management tools discussed later will help develop contingency plans so that any delays can be avoided.

Managing the project using all the tools described includes recognizing any conflicts in resource availability. The resources include tooling,

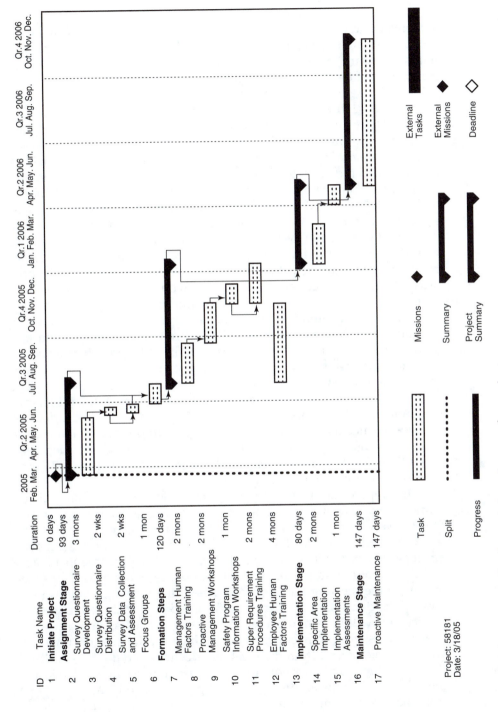

FIGURE 8–1 The four major phases of a project

personnel, and materials. As activities are completed, the project manager must assess and recalculate the critical path. Any revisions or overtime hours necessary must be managed quickly and effectively to guarantee completion of the project on time and under budget.

Managing the team A vital ingredient to the effective performance of any project is successful team management. This includes managing the internal and external boundaries of the project itself. Internal boundaries include any exchanges between the project team members and the project manager. External boundaries are any communication between the project team or project manager and the customers or any stakeholders of the project. Stakeholders are defined as any person or group who has a "stake" in the project; that is, they benefit from the project completion. Managing the internal and external boundaries requires communication skills. Communication paths must be maintained between the project manager and the project team, between the project manager or project team and the customer, and between the project team and upper management. The communication process will go a long way toward determining whether a project is successful or not. The communication itself can be oral or written. One communication tool used by project managers is a progress meeting. The progress meeting is used to identify any problems, action items, or expectations from management that must be dealt with to keep the project on schedule. The project manager must prepare an agenda for each meeting and distribute copies to all participants. A project agenda is a document that describes the meeting objective, time, place, duration, participants, the items to be discussed in the meeting, who is responsible for each item, and a time estimate of each item on the agenda. (See Figure 8–2 for an example of a project meeting agenda.)

To prepare the meeting agenda, the project manager must determine the objective of the meeting. Meeting management can be crucial to the success of the project itself. Project participants are generally overworked, so effective and efficient meetings are crucial to help them manage their time better. The meeting agenda should be prepared in advance and distributed to project participants at least 24 hours before the meeting. After the meeting has been completed, the meeting minutes (a review of the discussion of the meeting) should be distributed to meeting participants within 24 hours after the meeting. A successful project manager handles project meetings with minimal arguments, on time, and with valuable output from the meeting. Meetings should be cancelled if there is nothing to discuss with the project team. A meeting called just for the sake of meeting can disrupt project progress and create resentment from the project team toward the project manager.

Agenda for Project Status Meeting

Meeting Objective: To inform the project team of accomplishments, project status, and identify any necessary corrective actions

Logistics	Meeting Members
Date:10/21/01	1. Leader: Edie
Time: 8:00-9:15	2. Attendees: Project Team
Location: War Room	3. Meeting called by: Edie 555-9292

Agenda Item	Process	Who Responsible	Time
1. Introduce agenda	Interaction/lecture	Edie	5 minutes
2. State meeting objective and give preview	Overhead	Edie	5 minutes
3. Identify accomplishments since last meeting	Gantt chart review	Steve	15 minutes
4. Review cost, schedule, and scope			
• Status	Project plan review	Matthew	10 minutes
• Potential problems		Jude	10 minutes
5. Corrective action necessary	Overhead	All	15 minutes
6. Action items identified	Interaction	All	15 minutes
7. Adjourn			

FIGURE 8–2   Sample agenda for a project progress meeting

8.3   MAINTENANCE OPERATIONS ANALYSIS

In this section we take a quick look at some of the maintenance activities that might evolve into longer-term maintenance projects. As in any critical operations area, maintenance planning is imperative for all areas of the facility to guarantee a first-class facility. This planning should include daily, weekly, monthly, and longer-term maintenance activities. As you can imagine, the type of maintenance required for the short term can differ greatly from long-term requirements. The following section examines an example of maintenance planning from an aluminum fabrication company that highlights different types of activities.

8.3.1 Maintenance Operations Procedures

First, an overall strategic look at the maintenance operation is taken. The organization must determine the overall purpose of the maintenance department, define the specific duties of the maintenance personnel, and identify which maintenance work areas they will manage. Further information such as specific job responsibilities, procedures, and record keeping also are identified (see Figure 8–3).

The overall strategic plan will give credibility to the maintenance organization. The goal is that facility maintenance will be properly done and performed regularly. The maintenance department should be proactive and not just "fight fires," reacting to all daily maintenance problems. Most facilities that relied on reactive maintenance can tell you the high costs, inadequate staffing, and morale problems their companies faced. When maintenance planning goes bad, not only do internal departments point fingers at the problem areas, but external customers also are unhappy because of due-date delivery problems and poor quality, both possibly caused by poor maintenance planning.

8.3.2 Regular Maintenance Tasks Identified

An overall strategy helps guide the maintenance plan. A company maintenance management plan identifies the daily maintenance activities necessary. This ongoing maintenance plan is the impetus for world-class production, creating a proactive maintenance plan for a company. Figure 8–4 shows the maintenance management program for the same aluminum fabricator used in our earlier example. As you can see, the purpose of the maintenance group is clearly defined, highlighting the expectations of the group and clearly identifying what will *not* be included in the maintenance plan.

The expectations of the maintenance personnel are clearly spelled out in the maintenance management program, making the communication process among maintenance personnel, maintenance management, and outside organizations simpler. The program also identifies the specific procedures for all types of maintenance problems, including equipment inspection maintenance problems as well as urgent maintenance problems.

After the operations procedures are clearly described, the maintenance management plan identifies how the maintenance documents will be controlled, how continuous improvements will be tracked, and how

---

1.0   **Purpose**
       This set of procedures has been developed for the understanding and control of the Maintenance Department.
       Scope
       This set of procedures pertains to all maintenance functions.

2.0   **Definitions**
       2.1   *Maintenance:* Pertains to all work for the upkeep of machinery, building, and mobile equipment.
       2.2   Affected Areas:
           • Specific work areas are identified here

3.0   **References**
       3.1   None

4.0   **Responsibilities**
       4.1   Engineering and Maintenance Supervisor: Responsible for maintaining and updating this manual.
       4.2   Maintenance Supervisor: Responsible for upholding the procedures and recommending changes for this manual.
       4.3   Skill Crafts: Responsible for following these procedures and recommending changes to the manual if and when necessary.

5.0   **Procedures**
       5.1   Refer to table of contents for references to specific procedures.

6.0   **Record Keeping**
       6.1   The Engineering and Maintenance Supervisor will be responsible for the retainment of this manual.
       6.2   Annual audit will be performed on the manual, procedures, and practices.

7.0   **Records**
       7.1   Records of the audit will be retained by the Engineering and Maintenance Supervisor for a period of five years.
       7.2   Training records

FIGURE 8–3   Maintenance operating procedures in an aluminum fabricating plant
(Source: Courtesy of Steve Olesak, maintenance engineer.)

performance evaluation will be described. A sample maintenance work request form is shown in Figure 8–5.

8.3.3   Implementing Specific Maintenance Activities

In some work environments, many maintenance activities are performed on a fairly regular basis. Even though these types of activities typically do not fall under the definition of *project* because they are performed

Purpose

The purpose of satellite maintenance is to better serve production by improving equipment availability in specific areas through faster response, higher focus of knowledge on machinery, better visibility, enhanced relations with production, and increased enablers for continuous improvement.

Expectations

Expectations of the maintenance person are, but are not limited to
1. Safe work practices
2. Professional conduct
3. Repair and improve equipment
4. Clean and tidy workplace
5. First response to issues in their area
6. Call for assistance when needed
7. Make routine calls at least once a day to all of the areas assigned
8. Have formal discussions with production supervision weekly
9. Prioritize work orders and work requests
10. Fill out maintenance section of work request
11. Post yellow copy of work order on appropriate board
12. Fill out maintenance section of log books
13. Assist in developing PMs for machinery assigned in area
14. Train and coach operators in performing production PMs
15. Perform maintenance PMs on a timely basis and document the job
16. Determine which spare parts are needed
17. Repair or rebuild parts if warranted
18. Discontinue parts that are no longer needed
19. Train maintenance personnel about equipment in area
20. Obtain control and update documentation for area equipment that includes
    - Drawings
    - Procedures
    - Manuals
21. Label equipment such as valves and settings
22. Daily communication with maintenance supervisor
23. Assist in continuous improvement

Work Orders

There are five types of work requests identified for maintenance to perform: (1) Breakdown, (2) Urgent repair, (3) Originated from preventive maintenance inspection, (4) Normal repair, and (5) Modification/continuous improvement. Safety work requests are still submitted in the existing system; however, they will be addressed as top priority.

Breakdown work requests will be communicated to the maintenance person verbally via two-way radio. The operator will fill out the top three lines of the Maintenance Breakdown Service Log Sheet located in the Maintenance Breakdown Service Log Book. The maintenance person will fill out the darkened areas after the job is completed or when the job was stopped. The purpose of this log is as an easy vehicle for cross communication between people during an emergency type situation. It also serves as a history file for a particular piece of equipment.

FIGURE 8–4   Sample maintenance management program for an aluminum
                  fabricator

Urgent repair request will be in written form. The requester (operator or other) will use the Maintenance Work Request form located in plastic holding boxes located at each cell. The operator will fill out the upper half of the form, then separate the sheets and post the canary copy of the request on the maintenance visual board, place the white copy of the request in the "Written Maintenance Work Request Drop-Off Box" located at the maintenance satellite outpost area, and contact maintenance. The maintenance person will then take the work request and prioritize and schedule the work. Once the work is completed, the maintenance person will fill in the remaining information (what was performed on the job) on both copies, white and canary, and state that the job is complete if completed. The canary copy is reposted and the white copy of the work request is placed in the "Completed Work Order Drop-Off Box." The work order will then be reviewed/processed on a weekly basis and then filed in the appropriate equipment file.

Originated from preventive maintenance inspection requests will be made in written form and submitted to maintenance like an Urgent repair request. However, if the PM inspection reveals a breakdown-type issue, the operator should contact the maintenance person immediately.

Normal repair requests will be in written form. The requester (operator or other) will use the Maintenance Work Request form located in plastic holding boxes located at each cell. The operator will fill out the upper half of the form, then separate the sheets and post the canary copy of the request on the maintenance visual board and place the white copy of the request in the "Written Maintenance Work Request Drop-Off Box" located at the maintenance satellite outpost area. The maintenance person will then take the work request and prioritize and schedule the work. Once the work is completed, the maintenance person will fill in the remaining information (what was performed on the job) on both copies, white and canary, and state that the job is complete if completed. The canary copy is reposted and the white copy of the work request is placed in the "Completed Work Order Drop-Off Box." The work order will then be reviewed/processed on a weekly basis and then filed in the appropriate equipment file.

Modification/Continuous Improvement (change, upgrade, addition, removal, convenience) requests will be in written form. However, the request must be approved by the process supervisor, with his/her signature on the work request, before it can be submitted. The requester (operator or other) will use the Maintenance Work Request form located in plastic holding boxes located at each cell. The operator will fill out the upper half of the form, then separate the sheets and post the canary copy of the request on the maintenance visual board, place the white copy of the request in the "Written Maintenance Work Request Drop-Off Box" located at the maintenance satellite outpost area, and contact maintenance. The maintenance person will then take the work request and prioritize and schedule the work. Once the work is completed, the maintenance person will fill in the remaining information (what was performed on the job) on both copies, white and canary, and state that the job is complete if completed. The canary copy is reposted and the white copy of the work request is placed in the "Completed Work Order Drop-Off Box." The work order will then be reviewed/processed on a weekly basis and then filed in the appropriate equipment file.

Communications

The satellite maintenance person will have available many forms of communication. Those vehicles include two-way radio, work orders, work order posting, and verbal. The maintenance person will check twice a day with each cell that they are responsible for to see how it is functioning.

(continued on next page)

FIGURE 8–4 *(continued)*

Equipment Inspections

Equipment inspections will be performed by both the operators and maintenance. Inspections can and will happen daily, weekly, monthly, quarterly, semiannually, and annually. The maintenance person will be responsible for developing and training operators in the proper techniques for machine inspection. Master copy of each inspection will be located in the Maintenance Supervisor's Equipment Files. The weekly/monthly/quarterly lubrication and preventive maintenance schedules are necessary to assure machine reliability, and safety. These procedures must be followed to discover any problems not normally seen in daily operations and assure that lubrication is adequate for machine operations. There will be in place a "mini-lubrication center" that will be maintained by the area responsible maintenance person. This area will have adequate grease guns and oil reserves.

Weekly on a day selected by the crews, the procedures outlined on the "Weekly Operator Inspection/Lubrication Procedure" will be checked and normal operating levels of fluids will be adjusted. Below each week (1–13) is a place for initials. Please place first and last name initial in this spot to assure that it was in fact checked. Mark in the appropriate box at the top of the sheet what quarter of the year it is and write the year in box to the right of the quarter area. If a noticeable amount of fluid was used, state in the comments box what was added where, and how much approximately was added. This will let the supervisor or area maintenance person be aware of a possible problem and start investigating within the week, so the problem is not noted again on the following week's inspection.

Monthly during the first week of the month the "Monthly Operator Inspection/Lubrication Procedure" will be checked out and filled out properly on supplied sheet. Again note any areas of concern in the "Comments" section at the bottom of the sheet.

As we quarterly have to inspect chains, slings, and lifting devices, so shall we quarterly perform a more extensive lubrication. This will not only include the weekly and monthly duties but will further insure that proper moving parts receive adequate lubrication. Please follow the steps on the quarterly lubrication list, and initial each item to assure it has been completed.

A looseleaf binder or similar book will be maintained at the machine so it can be accessed by crew, supervisor, or area-assigned maintenance person to check for any lubrication or maintenance concerns.

Thank you for your cooperation, and if we discover ways to make this more simple, while maintaining machine reliability, we can explore further upon request. Examples of PMs are attached (see Appendix XX). The maintenance person will retain all completed PM forms for a period of four quarters. Once a PM scheduled sheet is completed, the sheet should be delivered to the "Written Maintenance Work Request Drop-Off Box." The maintenance person will then file it under the PM section of the specific equipment's files.

Document Control

Document control falls under three categories: retaining, updating, and deleting documents. Retaining documents means to collect all useful information of the specific items and store them in a logical manner for easy accessibility at a later date. All information includes drawings, operators manuals, maintenance/service manuals, and equipment inspection documents. These documents will be located at the satellite maintenance area. The area's equipment will have individual files. These files will be subcategorized into completed work order requests, completed equipment inspections, blank equipment inspection sheets, spare parts inventory check sheets, and purchased requests.

FIGURE 8–4   *(continued)*

Continuous Improvement

The satellite maintenance mechanic will assist in continuous improvement. This includes working with production on issues, attending the continuous improvement meetings, and analyzing breakdowns for improvements.

Success Measurables

Success measurables are ways to effectively determine the success of the operation of service. The measurables will include but not be limited to
- Equipment downtime
- Work order backlog
- Response time
- Span of time equipment was down per breakdown
- PM Achievement Rate
- Total work orders performed

1. Equipment downtime—The time the cell is down for maintenance purposes other than scheduled maintenance both on a weekly and monthly bases.
2. Work order backlog—The number of work orders outstanding on a weekly basis.
3. Response time—The time from when maintenance is contacted until they show up at the machine.
4. Span of time equipment was down per breakdown—The measurement time between when maintenance is called and the job is completed.
5. PM Achievement Rate—The total PMs performed divided by the total PMs required on a monthly basis.
6. Total work orders performed—Number of work orders completed on a weekly and monthly basis.

After Hours

After-hour coverage of the areas will be performed by the general maintenance group. However, they have the responsibility of filling out the necessary paperwork (work orders) but will not be required to post the yellow copy of the work request on the cell's maintenance bulletin board. They will need to make entries in the log book and leave written work orders in the "Written Maintenance Work Request Drop-Off Box," where the satellite maintenance will review the work and then post the work request.

Vacations

Vacations will be filled by the general maintenance group. They will need to make entries in the log book and leave written work orders in the "Written Maintenance Work Request Drop-Off Box," where the maintenance supervisor will review the work and post the work requests.

Training

The maintenance satellite person will be a primary trainer for operators with regard to equipment inspection and minor adjustments or repair work.

FIGURE 8–4 *(continued)*

Maintenance Work Request   XXXXXXXXX
                                                         Request Number

Work Type (1):	Short description of work needed or problem:
Equipment Name:	What Part of the Equipment:
Date of Request:	Requested by (name):	Date Needed (2):
Approved By:		

Additional description, explanation or comments (4):

_____

_____

_____

If modification, improvement, or change, state why needed:

_____

_____

THIS SPACE FOR MAINTENANCE USE ONLY:

_____

_____

_____

_____

Completed By:_____   Date:_____

OFFICE USE

WORK ORDER #:   EQUIPMENT NO.

Notes: (1) Work Types 1-Breakdown (needed immediately)
                                2-Urgent repair (within 24 hours)
                                3-Originated from preventive maintenance inspection
                                4-Normal repair (not needed within 24 hours)
                                5-Modification, change, upgrade, addition, removal, convenience
                (2) If date is indicated, state the reason for the date.
                (3) Approval of supervisor is needed for type 5 work.
                (4) Attach any sketches, prints, or other details as necessary.

FIGURE 8–5   A sample maintenance work request form

1. Developed need for new piece of equipment.
2. Send concept to engineering for priority and request available funding.
3. Project assigned to task group, i.e., engineer(s) including Single Point Accountability (SPA).
4. Engineer examines problem to determine possible alternatives.
5. If new equipment is deemed necessary, engineer develops task force to determine needs
6. Once conceptual project is designed, safety and maintenance reviews are performed.
7. Equipment specification is developed. (Specification is based on specific and general, "boiler tag," requests.) The specification requires equipment details to be reviewed for ease of service, standardized parts, and foreseeable issues. (General spec. was developed through years of experience based on possible oversights.)
8. Specification sent to perspective suppliers.
9. Cost analysis developed to determine financial feasibility.
10. If financially feasible, a request for capital financing is submitted.
11. When request is approved, purchase orders will be submitted.
12. If necessary, project leader will visit manufacturing facility along with necessary plant associates for review of equipment. Photographs are taken as the equipment is built for additional review with larger plant body, i.e., maintenance and appropriate production personnel.
13. Preliminary spare parts review begins.
14. Site preparation is conducted beforehand to speed installation. Offline setup may occur to address issues before the unit is committed to production.
15. Instructions are given to maintenance for installation. Once installation is completed, safety review is again performed with production, maintenance, safety and engineering.
16. Lubrication charts are developed.
17. Safety alerts are generated, i.e., lockout procedures and hand safety analysis.
18. Training is performed for production, maintenance, and other stakeholders to the project.
19. An itemized list of the equipment's parts is developed and entered into spare parts database. The selection and securing of parts are performed.
20. Equipment is released to production.
21. Informal feedback meetings take place daily the first week and then the frequency is reduced to an as-needed basis.
22. Formal project review is performed and report is generated.

FIGURE 8–6   Project flowchart for design and installation of generic equipment at an aluminum fabricating company

repetitively, some project management techniques can be helpful tools for managing these important maintenance tasks. Figure 8–6 shows a sample project plan, identifying specific work tasks for the installation of generic equipment at the aluminum fabricating company discussed earlier.

8.3.4   Scheduling Work Items

Maintenance scheduling projects fall into two main categories: work items that must be completed in the upcoming period and long-term work items, which are part of a larger project effort.

The first category, work items that must be completed in the upcoming period, includes any emergency maintenance work, scheduled preventive maintenance, and daily housekeeping maintenance activities. The second category, long-term work items, includes capital projects, annual projects, and repair projects. These long-term work items are best managed using project management techniques because they involve multiple functions within the company.

Before determining what work items need to be scheduled in the upcoming period, there are some planning steps that must be taken. For routine preventive maintenance and daily housekeeping maintenance activities, it is likely that the planning already has been completed. Information needed to answer questions about what task to perform, what steps to complete, and how long the task will take should be available to maintenance personnel. For emergency maintenance work, these questions need to be addressed, along with identifying the scope of the job and who should perform the work.

Scheduling work items that must be completed in the upcoming period   After the planning of daily and routine maintenance jobs is complete, the next step is to prioritize overdue or upcoming jobs and emergency work. As in most production environments, an important factor is scheduling the maintenance work when the production equipment is not needed. If there is available production downtime, any maintenance work possible should be scheduled then. Any tooling and material needs must be assessed, and naturally, the workforce skills and availability should be evaluated. As shown in the aluminum fabrication company example, all maintenance work areas should be identified and scheduled at the proper time, with all available personnel. Naturally, all work items should be tracked to ensure their completion, and any disruptions to the production process should be communicated in advance of shutdown.

Scheduling maintenance jobs that require extensive planning, such as longer-term maintenance projects that require the use of project management techniques, will be discussed in Section 8.4, "Project Management Tools and Techniques."

8.3.5   Time Estimates

A key element in scheduling maintenance work is performed during the estimating process. Why are these estimates necessary? They are important for several reasons:

1. For longer-term maintenance projects, to see if the project is feasible
2. For future reference to compare time planned versus actual time expended

3. To evaluate other alternatives

4. To identify risks

To get better estimates, use one of the following methods:

1. Direct experience

2. Talk with someone with direct experience

3. Find a good model or example to use for comparison

4. Have someone explain the process

Obviously, the best estimate possible is obtained if the project manager or maintenance planner already has accomplished the work step before and thus has a good idea of how long the task will take. If you have not completed a comparable type of task before, try to find the experts in the area. Look for someone who is considered highly skilled in the process in question and get his or her input about how long the process should take. If you have the opportunity, do some research and identify good examples of similar processes. These often make a solid basis for a preliminary estimate. If no information about similar situations is available, ask the workers who perform the work step exactly what will take place and then talk through the specific work steps to make an estimate of the time required.

8.3.6 High-Level Estimates

The overall project estimates can be completed once the work steps and deliverables have been identified. This process includes estimating the maintenance effort, calculating costs, and estimating the elapsed time to complete the worksteps.

1. Estimate effort
   - Define estimating unit and level of detail
   - Estimate most likely, optimistic, and pessimistic effort
   - Identify assumptions that determine the estimates
   - Use high-level statistics to summarize work tasks and project estimates

Estimating is primarily a communication process. This assumptions/ estimating process improves communication early in the project.

For each work step in the project, estimate the most optimistic, most pessimistic, and most likely amount of effort the work step will take. We discuss this estimating process in greater detail in the latter part of the chapter. Project directors then can use statistics to improve the time estimates. For the most optimistic case estimate, identify a minimum amount of time that the project could take. Look at the assumptions for each work step to identify the most optimistic estimate and ask yourself, "What could go right?"

The most likely estimate is the amount of time that you would estimate the project would take if you could only choose one time estimate. Once again, identify the assumptions under which this time would be applicable.

The most pessimistic case estimate is the most time that the project could possibly take to complete. Evaluate the most pessimistic case estimate by identifying all the assumptions for "What could possibly go wrong?"

2. Calculate costs
   - Select method: average or specific rates
   - Select team and identify rates
   - Multiply rate by effort hours
   - Summarize for project cost

3. Estimate elapsed time, using effort-elapsed time conversion
   - Calculate impact of number of resources
   - Add duration for delays
   - Document assumptions made
   - Use high-level statistics to summarize subphase duration

This section has focused on some methods for managing shorter-term work items. Now we move on to the tools necessary for completing successful long-term maintenance projects.

8.4 PROJECT MANAGEMENT TOOLS AND TECHNIQUES

There are two popular project management network models with easily understood concepts yet proven effective real-life applications. These are the critical path method (CPM) and the program evaluation and review technique (PERT).

The study of Gantt charts may be enhanced by converting these bar-chart views of a project into a network diagram. The network approach extends the plot of the project into sequential and parallel tasks (activities) for a more enlightening overview and allows for the computation of pertinent time and cost factors. This network procedure is just an extension of Henry Ford's effective "exploding backwards" method of management from the early 1900s. That is, Ford visualized a completely assembled Model T automobile as the total of all its component parts. Then, the analysis "backed up," calculating when material orders and the assembly of each part (engine, body panels, wheels, etc.) had to begin so that all parts were on hand at the final assembly point.

The critical path method dates back to 1957, when it was developed by Remington Rand and DuPont for use in the construction and maintenance of DuPont chemical plants. The program evaluation and review technique was developed in 1958 by the U.S. Navy to plan and control the Polaris missile submarine program.

Steps in the conduct of both the CPM and the PERT network analyses are similar. The difference in the two approaches involves the estimated times to complete each task or "activity" of a project. CPM is deterministic, assuming that fixed-time estimates for the activities are sufficiently accurate. PERT is stochastic (probabilistic), eliciting probability distributions for the task times.

These models are versatile. That is, they are useful for several functions of management, and can be used by different staffs within the firm. You are probably familiar with lists of "the functions of management," for which the terminology has changed again and again over the years. But typically, *planning, organizing, scheduling, controlling*, and so on are listed as the functions of management. Many excellent management science and operations management texts show how different analysis models (linear programming, waiting lines, transportation and assignment, inventory, etc.) are applicable only to certain functions within the company. In contrast, network models can be constructed in the planning and scheduling of a project and then can be used to control subsequent project implementation and required adjustments to the initial plan.

The number of experienced managers who have never been exposed to many of the key course topics in quantative analysis courses is always rather astounding. Such students become enthusiastic about the applicability of these "new" techniques to their responsibilities within the firm.

One mature student, a high-level manager with an international firm in South America, was taking a break from his job to return to graduate international business studies. His previous education had not included the most likely/optimistic/pessimistic time (or cost) estimates used in the PERT procedure, which is covered later in this chapter.

This manager, when shown the three time estimates procedure, hit himself on the forehead and said, "Wow! This technique really would have helped me in my job!" He had needed quarterly estimates from his area managers for critical and rapidly changing factors, such as rates of inflation. He had asked for fixed value estimates, which clearly stressed the area managers. Though experienced in their jobs, they were hesitant to commit to specific values. He realized that the technique of requiring most likely, optimistic, and pessimistic estimates would have resulted in much more accurate factor estimates from his area managers. He vowed to incorporate this procedure when he returned to South America.

Simple project examples will be used to demonstrate pertinent network model procedures. The reader should be confident that these procedures effectively apply to real-life, complex projects of many more tasks than the text illustrates. Fortunately, many excellent computer software programs are available to help the manager solve complex network models at the click of an icon. But, for meaningful analyses, the manager should understand the theory and concepts involved in the model being used, before depending on push-button computer solutions.

8.4.1   Basics of the CPM and PERT Networks

In showing the project management network model, we'll start at the beginning of Henry Ford's "exploding backwards" concept and logically work through all related tasks to the end point.

1. Identify all significant activities involved in the project.

2. Develop the logical sequential and parallel relationships among the activities. A project activity table will define this rationale.

3. Depict the time relationships on a network diagram.

4. Annotate activity time estimates on the diagram.

5. Compute the longest ("critical") path of activities through the network.

6. *Use* the project network scheme and related factor computations to plan, schedule, monitor, and control the project times and costs throughout its implementation. Periodic adjustments may be necessary as activity times vary from the initial estimates.

Identifying project activities and their relationships   To illustrate the first two steps of the network concept, let's define a simple project involving construction of an automobile component manufacturing plant. The plant might manufacture alternators or radiators—*any* component will serve our purpose. In demonstrating the network diagram, the common convention of "activity-on-arrow" and "event nodes" will be used. At the end of the chapter, another convention, "activity-on-node," will be shown. Most of the current computer software network programs accept data input in either format. Activity arrows will be given letter notation, and event nodes will be numbered. A versatile project activity table for the example is shown in Table 8–1. The reader will see that the Event Node sequence and the Immediate Predecessors columns define the same network relationships, but the redundancy is useful in giving the analyst a clear visualization of the project makeup and how the subsequent network diagram will look.

TABLE 8–1

Project Activity Table for Constructing a Manufacturing Plant

Activity	Event nodes	Description	Immediate predecessors
A	1–3	Zoning Application	—
B	1–2	Plant Design	—
Dummy	2–3		B
C	3–4	Building Permit	A, Dummy
D	4–7	Plant Construction	C
E	4–5	Machinery Orders	C
F	5–7	Machinery Installation	E
G	4–6	Employee Hiring	C
H	6–7	Employee Training	G
I	7–8	Operational Testing	D, F, H

It should be clear that event node 1 is the *Start* node, and node 8 is the *End* or *Complete* node. All activities starting at the same number node have as their immediate predecessor(s) any activity that ends at that numbered node.

Drawing the network diagram   Now look at the project network diagram that is described by the given activity table, shown in Figure 8–7. This illustration can be used to explain some important conventions of properly drawn networks. Letter labels are placed under each activity arrow. Dummy activities typically are drawn as dashed lines for clarity.

Let's explain some standard conventions for the construction of PERT and CPM network diagrams, as demonstrated in Figure 8–7. The sequential versus parallel activity logic should be clear. That is, A and B,

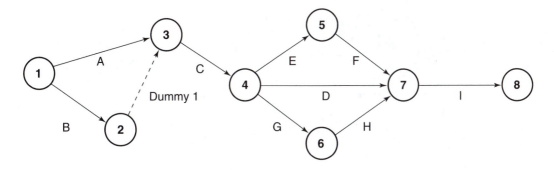

FIGURE 8–7   Network for constructing a manufacturing plant

Zoning Application and Plant Design, can proceed at the same time; they are parallel. However, C, Building Permit processing, cannot begin until the proper zoning is confirmed *and* the specified building design has been approved. After C is completed, three subsequent activities, D, E, and G, can begin at the same time and can be accomplished in parallel.

Activity arrow directions indicate the progress in time. The *length* of the arrows merely permit convenient drawing of the network activity relationships. Activity *times* are not comparable to the scale of the arrow lengths. A shorter arrow activity may take longer to complete than a longer arrow activity; the network diagram shows the relationships between activities, but it is not a project time line.

Activity arrows are shown as straight lines. That practice sometimes results in the need to use "dummy" activity arrows, but the resulting clarity of the network is better than if the straight arrow convention were violated. It is tempting to show activity A and B relationships by two *curved* lines from node 1 to node 2, but that would be incorrect. Figure 8–8 shows a network diagram drawn improperly with curved lines connecting the nodes.

Again, when several activities *end* at one event node, that indicates that no activity *starting* at this node can begin until all activities ending at the node are completed.

Assigning activity time estimates   For the purpose of showing network computations leading to determining project time and the critical path, let's assign fixed or deterministic time values to each activity, as in the CPM procedure. These activity time estimates are provided by the

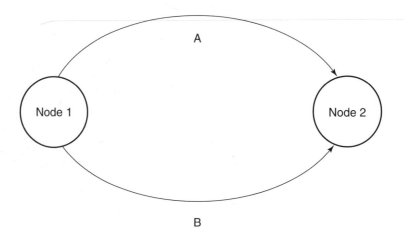

FIGURE 8–8   Sample of an incorrect drawing of a network diagram

TABLE 8–2

Project Activity Table with Time Estimates Added

Activity	Nodes	Activity time (weeks)	Immediate predecessors
A	1–3	12	—
B	1–2	10	—
Dummy*	2–3	0	B
C	3–4	3	A, Dummy
D	4–7	23	C
E	4–5	17	C
F	5–7	8	E
G	4–6	12	C
H	6–7	8	G
I	7–8	5	D, F, H

* If this activity table were entered into existing PERT or CPM computer software programs, the dummy activity could be omitted; activity C could be shown with *two* immediate predecessors, A and B. The computer program would recognize the correct relationship and *add* the necessary dummy activity before it printed an activity-on-arrow chart.

responsible task managers who, with experience, should be able to give accurate time forecasts. Table 8–2 adds activity time estimates in weeks to the data in Table 8–1.

We might annotate activity times as small *t*s. By that convention, t_a is 12, t_b is 10, and so on. The project completion time will then be given the notation of capital T.

Now the procedure is to track through the network to find how many weeks the project will require, and which sequence of activities form the *critical path*(s) for the total time. This is accomplished by doing a "forward pass" to determine each activity's *earliest start* (ES) and *earliest finish* (EF) times, starting at node 1 and ending at node 8. The project time will have been determined after completing the forward pass. Then, a "backward pass" will determine *latest finish* (LF) and *latest start* (LS) times by tracking backward from node 8 to node 1. Following are definitions of these terms to facilitate understanding before we proceed to the actual forward/backward operations:

> *Earliest start time* for an activity starting at a node is the largest *earliest finish time* for all activities ending at that node. In other words, subsequent activities cannot start until all predecessor activities have been completed.

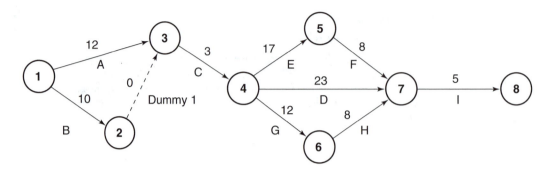

FIGURE 8–9   Plant construction network diagram with activity times

Latest finish time for an activity ending at a node is the smallest *latest starting time* of all activities starting at that node. In other words, prior activities must be completed in time so as not to delay the earliest possible start of subsequent activities (or the project will be delayed beyond its shortest possible completion time).

Figure 8–9 shows the same network diagram as in Figure 8–7, now with the activity time estimates (in weeks) above each activity arrow. This basic diagram will facilitate our forward and backward pass calculations of ES, EF, LF and LS times.

As the calculations proceed, each activity's ES and EF times will be shown at the beginning and end of the arrow, *above* the arrow. Later, when the backward operation proceeds, LF and LS times will be put at the end and beginning of each activity arrow, *below* the arrow. Do these operations manually to understand the concept before exploiting computer software solutions for the same project. Use different colors for forward and backward pass times; for example, use blue for ES and EF times and red for LF and LS times to clearly see the analysis computations and avoid calculation errors caused by misreading a "busy" network diagram.

Now let's start the forward pass. Clearly the earliest start time out of node 1 for both activities A and B is 0. Thus A's earliest finish time is 0 + 12 = 12 at node 3. B's earliest finish time is 0 + 10 = 10 at node 2. Then carry B's EF time around node 2 to serve as Dummy's ES time. Remember, any activity's ES time out of a node is the largest of all predecessor activity EF time into that node. Now Dummy's EF time is 10 + 0 = 10 at node 3.

Now we have *two* EF times ending at node 3. Therefore, activity C's ES time is 12 weeks because *all* predecessor activities into its starting node must be completed before activity C can start. That is, the largest EF time, A's 12 weeks, is C's ES time.

Continue on through the network. C's EF time is $12 + 3 = 15$ weeks. It is the only activity ending at node 4, so all activities starting at that node, E, D, and G, have ES times of 15 weeks. Then:

$$\text{E's EF time at node 5 is } 15 + 17 = 32 \text{ weeks}$$

$$\text{D's EF time at node 7 is } 15 + 23 = 38 \text{ weeks}$$

$$\text{G's EF time at node 6 is } 15 + 12 = 27 \text{ weeks}$$

Activity F's start must wait for the completion of E, so F's ES time is 32; its EF time is $32 + 8 = 40$ weeks at node 7. Activity H's start must wait for completion of activity G, so its ES time is 27; its EF time is $27 + 8 = 35$ weeks at node 7.

There are now three EF times ending at node 7 (F = 40, D = 38, H = 35). The start of activity I must wait until the largest of these (F = 40) is completed. Therefore, I's ES time is 40 and its EF is $40 + 5 = 45$ weeks at node 8, the *end* node for the project.

The completed forward pass provides the largest EF time at the end node, 45, and this *is* the project time. So $T = 45$ weeks, based on our best planning estimates; that is, the manufacturing plant should be operational 45 weeks from the start of the effort. Figure 8–10 shows the same diagram as in Figures 8–7 and 8–9, updated with the calculated activity ES and EF times above the arrows, preferably, as noted previously, using one specific color.

But which activity sequence forms the "critical path"? That is the path along which the project time is in a "bind." Delay of any activity along that path will delay the entire project. There is no slack time on the critical path. To calculate LF and LS times that will identify the critical

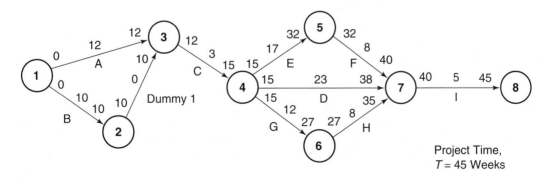

FIGURE 8–10   Plant construction network activity ES and EF times

path, we now complete an analysis backwards through the network, starting with the project time set as the LF time for all activities ending at the last node (45 weeks in this case). Work backward, then, calculating each activity's latest start time (LS), which is its LF time minus the estimated activity time.

The first step of the backward pass sets activity I's LS time as $45 - 5 = 40$ weeks at node 7. This is then the LF time for all activities ending at node 7 (F, D, and H). If they do not finish by the 40-week point, the project will be delayed because activity I cannot begin on time. The LF time for an activity ending at a node is the *smallest* LS time of all activities starting at that node. Do the calculations:

For F, LS time out of node 5 is $40 - 8 = 32$ (This is activity E's LF time)

For D, LS time out of node 4 is $40 - 23 = 17$

For H, LS time out of node 6 is $40 - 8 = 32$ (This is activity G's LF time)

Continuing backwards:

E's LS time out of node 4 is $32 - 17 = 15$ weeks

G's LS time out of node 4 is $32 - 12 = 20$ weeks

Note now that there are three differing LS times out of node 4; E = 15, D = 17, and G = 20. To prevent any project delay, all activities ending at that node (in this case just one, activity C) must finish to meet the earliest LS time out of that node. Therefore, activity C's LF time is 15 weeks, and its LS time is $15 - 3 = 12$ weeks.

The LF of all activities ending at node 3 must meet the smallest LS time out of node 3, which is C's 12 weeks. So the LF times of A and Dummy are 12 weeks:

A's LS time is $12 - 12 = 0$

Dummy's LS time is $12 - 0 = 12$, which is then B's LF time, and

B's LS time is $12 - 10 = 2$

The upgraded network diagram including both forward and backward pass computations (ES, EF, LS, and LF times) is shown in Figure 8–11.

You should have noticed the *critical path* early in the backward pass procedure. That is the route along which the project has no slack time, the path along which activity LS times are the same as ES times, and EF and LF times also match. This is clearly path A → C → E → F → I or nodes 1 → 3 → 4 → 5 → 7 → 8. In Figure 8–11, the nodes along the critical path are shaded. You might darken the activity arrows along this path to

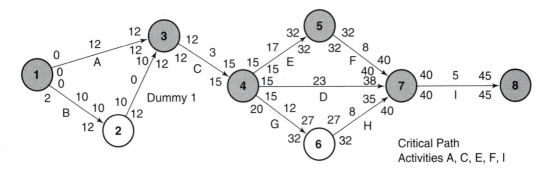

FIGURE 8–11   Completed plant construction network (ES, EF, LS, and LF times added)

highlight the critical sequence of project activities. Those, of course, deserve management priority to preclude project delay. Other paths have slack time. For example, activity H can start as early as the 27-week point. But it *need not* begin until week 32—it has five weeks' slack.

The tabular presentation of the completed network analysis is presented in Table 8–3.

Slack times provide the manager flexibility in the alternate application of unused resources. The eight-week training activity (H) with five weeks' slack permits the training instructors to work *any* eight-week period between the 27th and 40th weeks. They can be utilized elsewhere in the firm during their five-week slack time on *this* project.

TABLE 8–3

Project Analysis (Forward and Backward Pass) Results

Activity	Nodes	t	ES	EF	LS	LF	Slack
A	1–3	12	0	12	0	12	0*
B	1–2	10	0	10	2	12	2
Dummy	2–3	0	10	10	12	12	2
C	3–4	3	12	15	12	15	0*
D	4–7	23	15	38	17	40	2
E	4–5	17	15	32	15	32	0*
F	5–7	8	32	40	32	40	0*
G	4–6	12	15	27	20	32	5
H	6–7	8	27	35	32	40	5
I	7–8	5	40	45	40	45	0*

*Critical path of binding activities.

Activity B plant design would certainly start at time 0, knowing that it has two weeks' slack while waiting on the 12-week zoning application process. Perhaps zoning might not require the full 12 weeks, or maybe design might take longer than 10 weeks for added refinements. In any case, recognize the comfort of slack activities as opposed to the urgency of the binding tasks.

8.4.2   PERT Network Probability Analysis

In real life, project managers seldom (probably *never*) have the luxury of precise fixed-time estimates for project tasks. This statement is not made to denigrate the usefulness of a project overview using deterministic estimates, however.

The chapter introduction alluded to a project manager's difficulty in obtaining deterministic factor values from area managers. A three-time-estimate technique permits task managers to comfortably input values for factors within their areas of expertise. Let's refer to *time* estimates here, although *cost* or *weather conditions* or various other project factors can be similarly estimated.

Activity and project time and variance computations   The PERT probabilistic method extracts three time estimates for each activity. Managers experienced in their areas of responsibility can think through accurate values for

a = Optimistic (shortest) time to complete an activity

m = Most likely time required to complete an activity

b = Pessimistic (longest or worst-case) time for an activity completion

You may have experienced a project on which "everything went perfectly." For example, ideal weather, on-time material deliveries, and a motivated and experienced work force on a construction project all would work together to make the job move smoothly and quickly toward completion. On the other hand, there could have been repetitive storms, material delays or quality problems, worker illnesses, and other conditions that created an "everything went wrong" result. The experienced supervisor or foreman from whom the three time estimates are requested can factor in related variables for both types of situations.

From research and from proven applications, we can assume that an activity time has a *beta* probability distribution. There have been occasional criticisms of this assumption, but the beta distribution has been successfully used in PERT applications for many years.

For a beta-distributed activity time for which the three time estimates have been elicited, its average or mean value is calculated as

$$t = \frac{a + 4m + b}{6}$$

and the activity time variance, v (or σ^2 in standard notation), is

$$v = \frac{(b - a)^2}{36}$$

If the *project* time then can be calculated as a variable, with a defined probability distribution, the manager has much more *information* than when using deterministic times that will seldom occur in practice. That is, he or she is given the capability to make confident statements about the possibility of different project times (or costs).

To illustrate, let's redo our example project using PERT time estimates, as shown in Table 8–4. The optimistic, most likely, and pessimistic time estimates are contrived to yield average times that are the same as the original deterministic time estimates.

Retaining activity variance values in fractions when summing avoids loss of accuracy in the computation of project variance. Then, the project figure might be rounded to decimal places, $248/36 = 6.88888$, or 6.889 would be sufficient accuracy.

TABLE 8–4
PERT Time Estimates for Plant Construction Example

Activity	a	m	b	t	v
A	9	12	15	12*	36/36 = 1.000
B	8	9	16	10	64/36 = 1.778
Dummy	0	0	0	0	0
C	1	3	5	3*	16/36 = .444
D	20	22	30	23	100/36 = 2.778
E	9	18	21	17*	144/36 = 4.000
F	6	8	10	8*	16/36 = .444
G	11	11.5	15	12	16/36 = .444
H	7.5	8	8.5	8	1/36 = .028
I	2	5	8	5*	36/36 = 1.000

*Critical path activities, totaling $T = 45$ weeks

Project time variance, summed along critical path $= 248/36 = 6.89$

Using these average activity times, the forward and backward pass results are the same as in the previous section, and *average* project time is 45 weeks.

Project time *variance* (sum up activity variances along the critical path), or *sigma-squared* in standard statistics notation, is $248/36 = 6.88888$. Remember from statistics that when you sum a series of distributions to form a new distribution, the summary variable mean is the sum of the single distribution means. But, the standard deviation of the summary distribution is the *square root of the sum of the individual distributions' variances.* You cannot add the individual standard deviations; you add the variances and then take the square root of that sum to calculate the summary variable's standard deviation.

The sum of a series of *beta* distributions is a *normal* distribution, also termed the *Gaussian* distribution, or the well known bell curve. Thus, the average project time

$T = 45$ weeks, with a standard deviation of

σ = square root of $(248/36)$, which is $2.62466 = 2.625$ weeks

Project time probability assessments   Visualize the project time T distribution as a normally distributed bell-shaped curve, as shown in Figure 8–12.

Now the project manager can derive much more project *information* than was possible using the fixed-value method. It is a given that the probability of exceeding 45 weeks project time is .5000 (half the area under the curve). Similarly, the probability of completing the project in *less*

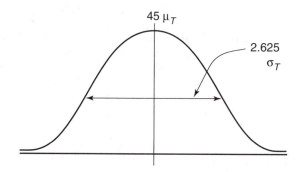

FIGURE 8–12   A bell curve, $T = 45$ at midpoint, σ of 2.625 annotated as distance from that center line to the points of inflection above and below the mean

than 45 weeks is .5000. But suppose a less-than-48-week goal was important to the firm. What is the probability of meeting or exceeding that goal (i.e., $P\{T < = 48 \text{ weeks}\}$)? Well, remember that normal-table calculation from statistics?

$$Z = \frac{X - \mu}{\sigma}$$

In this case, $Z48 = (48 - 45)/2.625 = 3/2.625 = 1.14$ standard deviations. And, from any quantitative text's normal distribution table, the area (probability) contained within 1.14 standard deviations is .37286. Therefore, the probability of completing the project in 48 weeks or less is greater than 87%. That probably would be satisfactory to the company's goal setters.

As a worst-case concern situation, suppose that a project time of over 50 weeks is totally unacceptable, given alternate-use plans for the same resources. Then $Z50 = (50 - 45)/2.625 = 1.90\ \sigma$, enclosing 0.47128 area, so the probability of exceeding $T = 50$ weeks is $0.5000 - 0.47128 = 0.02872$. So there is less than a 3% probability of the unacceptable outcome, which also should be comforting to the people who must make the decision to go ahead with the project or not.

One final probability assessment example examines how likely the project will be completed early, say, within 43 weeks. Then $Z43 = (43 - 45)/2.625 = -2/2.625 = -.762$, which contains about .28 area to the left of the mean. The probability of finishing in 43 weeks or less is $.5000 - .28$, or 22%.

Additional comments on PERT network analysis   Given a realistically complex project, it can have more than one critical path. Some of these paths may have common activities over part of their lengths, and then may diverge into parallel paths. In this case, the analyst calculates the project time (normal) distribution along each critical path. All the paths will have the same estimated project *time*, but variances usually will be different on each critical path. The critical path(s) of *most concern* then will be the one(s) with the greatest variance (relate this to the paths of *least control*). The project manager must pay attention to all critical paths as the project progresses and the network analyses are periodically upgraded. But the path(s) of greatest variance will be the controlling one(s) in the project probability analyses.

Time/cost tradeoffs   The manager can assess time/cost tradeoffs by analyzing how much additional cost is incurred if the project is reduced to less than the original time shown by network computations. Reduction in project time is a common requirement for many reasons.

For example, a company may have another important project scheduled to begin *after* this project. That effort may need the personnel resources currently committed to the first project, or perhaps the product of the first project is required as a subassembly for the second.

Activity time estimates are often given in terms of "normal" and "minimum" times with related costs so that time/cost comparisons are possible. Of course, the minimum time estimate is achieved at a cost penalty, requiring more resources or accelerated vendor material input and so on. We will look at a different project for this example, one that shows an activity table format that includes two-times/two-costs estimates. An assumption of linear per-period crash costs will be made. That is, suppose activity X has a normal time of 10 weeks at a cost of $12,000, and a minimum (full crash) time of seven weeks at a cost of $18,000. Assume each reduced week's cost is $2,000 (the $6,000 difference divided by three weeks). This linear relationship may be unrealistic; the first period reduction may cost less than subsequent one-period reductions. Increasing cost-per-additional-time-savings can be estimated and incorporated in the same time/cost analyses shown here. The linear assumption merely facilitates demonstration of the procedure.

Again, excellent computer software exists that allows the analyst simply to input the initial project activity table of precedents and normal/minimum activity times and related costs. The analyst then can specify the desired reduced project time and the software will solve this problem while printing out whatever analysis results are requested, such as solution tables, network diagrams, Gantt charts, and so on. All outputs are available for normal and crashed project times and costs. Many available management science computer packages also are capable of plotting related time/cost tradeoff graphs.

8.4.3 CPM/Crash Analysis: An Illustration

Table 8–5 shows an eight-activity example project in a normal/minimum time and related cost format. Then, Figure 8–13a depicts the network diagram for this project. Now both normal and minimum times are annotated above the activity arrows to facilitate accurate repetitive forward and backward pass operations. The analysis guideline coils for completing the project in 45 weeks' time at the minimum total project cost. Figure 8–13a will be upgraded as the time/cost analyses are iterated. These multiple network analyses could be quite tedious on a CPM-crash project of realistic size, which is why computers are used for these analyses.

The first forward/backward passes result in a 49-week project time at a cost of $54,000 (the sum of the Normal Cost column entries in Table 8–5). The critical path follows activities B → E → F → G.

TABLE 8–5

CPM-Crash Time/Cost Tradeoff Example

Activity	Immediate predecessor	Activity Time (Weeks)		Activity Cost ($1,000)		Cost/ week
		Normal	Minimum	Normal	Minimum	
A	—	10	8	9	15	3
B	—	13	10	7	13	2
C	A	15	13	3	5	1
D	A	10	10	7	7	—
E	B	8	7	10	14	4
F	D, E	8	7	7	11	4
G	C, F	20	19	6	12	6
H	D, E	9	6	5	8	1

To reduce the project time by one week, the reduction must be made along the critical (binding) path. The most economical one-week reduction for B, E, F, or G is B, at $2,000. Reduce B to 12 weeks, and do another forward/backward pass. The result of this analysis is shown in Figure 8–13b.

The second forward/backward pass operation illustrated in Figure 8–13b shows *parallel* critical paths: A → D → F → G and B → E → F → G. Notice that the next one-week reduction can be taken at F or G, or at *either* A or D (D cannot be reduced) in combination with *either* B or E. The additional cost of the one-week reduction would be $4,000 for F, $6,000 for G, $5,000 total for A *and* B, or $7,000 total for A *and* E. The optimal (minimum cost) decision will reduce F to its minimum time of seven

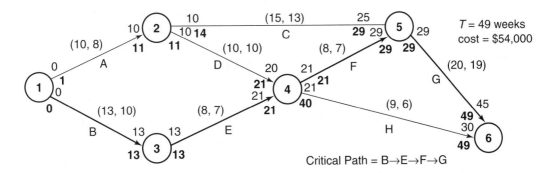

FIGURE 8–13a Network diagram for CPM-crash example (normal time ES, EF, LS, LF shown)

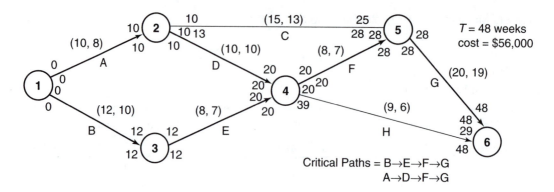

FIGURE 8–13b Diagram with forward/backward pass after activity B has been reduced

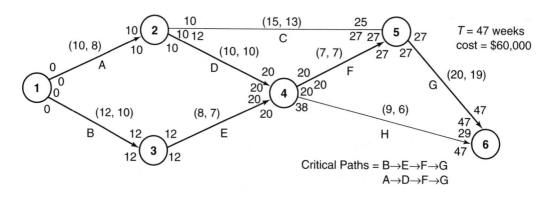

FIGURE 8–13c Diagram with both activities B and F now reduced by one week

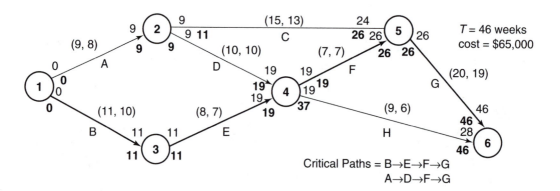

FIGURE 8–13d Diagram with activities A and B each reduced by one week (in addition to the previous one-week reductions in B and F)

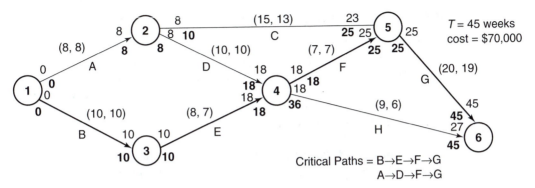

*Now activity G is the only activity on the critical path with available time reduction.

FIGURE 8–13e Diagram showing fourth and final forward/backward pass*

weeks. That will result in a project time of 47 weeks, and total cost now is $56,000 + $4,000 = $60,000. This diagram is shown in Figure 8–13c.

After the third forward/backward pass operations, the critical paths remain A → D → F → G and B → E → F → G. Activity F has been reduced to its minimum seven weeks, so the next one-week reduction options are G, at a cost of $6,000; A *and* B, at a total cost of $5,000; or A and E at a total cost of $7,000. The optimum decision reduces A to 9 weeks and B to 11 weeks. Project time is now 46 weeks, at a total cost of $60,000 + $5,000 = $65,000. That diagram is shown in Figure 8–13d. We still have to reduce the project by one more week to meet the management goal stated at the outset of this example.

The fourth and final forward/backward pass (Figure 8–13e) shows the same parallel critical paths as the last operation, and the one-week reduction candidates remain the same: G, $6,000; A *and* B, $5,000; or A *and* E, $7,000. Reducing A *and* B is again the optimal decision. They are reduced to their minimum times of 8 and 10 weeks respectively, at an additional cost of $5,000. Our goal has been achieved—project time, T = 45 weeks, at a total cost of $65,000 + $5,000 = $70,000.

If further project time reduction analyses were pursued, the analyst would find that the minimum possible project time is 44 weeks, at a cost of $76,000. This is achieved by reducing G to its minimum 19 weeks, at an additional cost of $6,000. *Note:* All activity times need not be reduced to their minimum to attain the minimum project time (that "full crash" would cost $85,000!). Minimum project time usually is reached long before full crashing. In this case, at a minimum project time of 44 weeks, all *critical* path activities A, B, D, F, and G are at their minimum times. Any additional activity time reductions (of C, E, or H) would not only waste money, but also would not affect the minimum time along the critical path(s).

8.4.4   A Realistic (Complex) PERT Project Example

Let's think through a realistically sized project. Specific activity descriptions are omitted, as they are not involved in the example project's mathematical analyses. Table 8–6 shows the project's activity table. Including the two dummy activities, there are a total of 26 activities. As mentioned in Table 8–2, the dummy activities need not be included in a computer program input—more on that point later.

The activity table and network diagram   Several points are pertinent here. First, you should appreciate how difficult it is to translate the

TABLE 8–6

Activity Table for Large Project Example (BIGEX)*

Activity number	Activity name	Immediate predecessor (list number or name, separated by ' , ')	Optimistic time (a)	Most likely time (m)	Pessimistic time (b)
1	A		4	5	12
2	B		1	3	5
3	C		2	7.5	10
4	D	A	6	8	10
5	E	A	2	3	10
6	F	A,B	2	4	12
7	G	A,B	9	10	17
8	H	C,F	5	7	9
9	I	C,F	7	9	11
10	J	D	8	11	20
11	K	E	4	6	8
12	L	G,H	2	4	6
13	M	G,H	6	15	18
14	N	I	9	11	13
15	O	I	4	12	14
16	P	I	13	15	17
17	Q	J,K,L	4	6.5	18
18	R	M,N	1	2	3
19	S	M,N	8	8.5	12
20	T	Q,R	4	6	8
21	U	O,S	8	10	24
22	V	O,P,S	10	18	20
23	W	T	3	3.5	7
24	X	U	9	10	17

*The network diagram (activity-on-arrow convention) is shown in Figure 8–14.

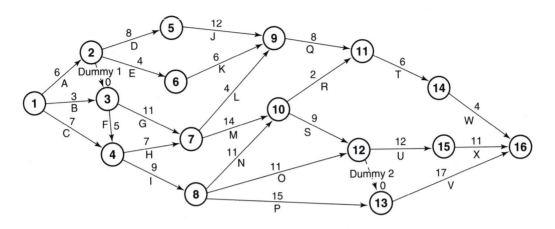

FIGURE 8–14   Network diagram for large project example

project activity table into the related network diagram. It is a tedious draw-erase-redraw-erase-redraw exercise until the diagram is correctly and clearly completed, without *crossing* any activity arrows if at all possible. Without referring to Figure 8–14, try to transition from Table 8–6 to Figure 8–14 on your own.

A second point: the average activity times have been calculated and are shown on the network diagram. You can confirm the accuracy of those by using the following equation for each activity:

$$T = \frac{a + 4m + b}{6}$$

It should be clear by now that some activity time estimates are symmetrical about the most-likely value; others are "skewed" in one direction or another. But these three-value estimates do represent the best efforts of experienced foremen and supervisors to provide accurate data.

Also, the average project time T, more accurately annotated as "μ_T" in a probability environment, is 64 weeks, over a single critical path of activities A → Dummy1 → F → H → M → S → U → X. The project variance on that path, call it "$\sigma^2{}_T$," is 18.333 weeks. The square root of that gives a project standard deviation of $\sigma_T = 4.2817$ weeks. A forward/backward pass confirms the given critical path sequence and project variance computation along that path. Again, these suggested manual analyses are aimed at giving the reader *full* appreciation for the analysis efficiencies of

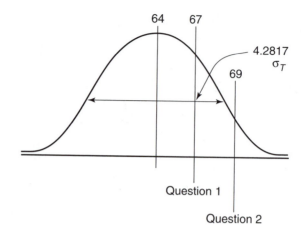

FIGURE 8–15 Example project average time distribution

current network model computer software. One such application software type will be discussed shortly.

Some project management concern computations Now we will
work through a few probability analyses, and then we will look at a specific software program's corollary solution. We know the average project
time is 64 weeks, with a standard deviation of 4.2817 weeks. The project
then can be visualized as the normal distribution shown in Figure 8–15, a
bell curve with its midpoint at 64 and standard deviation distance from
midpoint shown as ±4.2817.

Calculate the answers to these questions of concern to project
management:

Question 1: What is the probability of completing the project in 67
weeks or less?

Question 2: What is the probability that the project time will be
over 69 weeks?

Solution to Question 1 Z value is $(67 - 64)/4.2817 = .7$, which encloses an area of .2580 above the mean. Thus, the probability of completing the project in 67 weeks or less is .5000 + .2580, or .7580, or about 76%.

Solution to Question 2 Z value is $(69 - 64)/4.2817 = 1.168$, a value
that encloses an area of approximately .3790. The probability of the project time exceeding 69 weeks is .5000 − .3790 = .1210, or about 12%.

8.5   COMPUTER PROGRAM SOLUTIONS

As a preface to subsequent computer program printouts for previous example projects in this chapter, Figures 8–16a through 8–16e show a series of input/output printouts for our current example. The filename "BIGEX" was specified for these runs. They are done on the popular Quantitative Systems for Business (QSB) program, specifically a Windows version.* QSB is frequently updated, and is in use at many universities.

Note that the solutions presented in Figures 8–16d and 8–16e are more accurate than our rounded-value example computations. Also,

Activity Analysis for BIGEX

05-26-2005 08:36:23	Activity Name	On Critical Path	Activity Mean Time	Earliest Start	Earliest Finish	Latest Start	Latest Finish	Slack (LS-ES)	Activity Time Distribution	Standard Deviation
1	A	Yes	6	0	6	0	6	0	3-Time estimate	1.3333
2	B	no	3	0	3	3	6	3	3-Time estimate	0.6667
3	C	no	7	0	7	4	11	4	3-Time estimate	1.3333
4	D	no	8	6	14	26	34	20	3-Time estimate	0.6667
5	E	no	4	6	10	36	40	30	3-Time estimate	1.3333
6	F	Yes	5	6	11	6	11	0	3-Time estimate	1.6667
7	G	no	11	6	17	7	18	1	3-Time estimate	1.3333
8	H	Yes	7	11	18	11	18	0	3-Time estimate	0.6667
9	I	no	9	11	20	12	21	1	3-Time estimate	0.6667
10	J	no	12	14	26	34	46	20	3-Time estimate	2
11	K	no	6	10	16	40	46	30	3-Time estimate	0.6667
12	L	no	4	18	22	42	46	24	3-Time estimate	0.6667
13	M	Yes	14	18	32	18	32	0	3-Time estimate	2
14	N	no	11	20	31	21	32	1	3-Time estimate	0.6667
15	O	no	11	20	31	30	41	10	3-Time estimate	1.6667
16	P	no	15	20	35	32	47	12	3-Time estimate	0.6667
17	Q	no	8	26	34	46	54	20	3-Time estimate	2.3333
18	R	no	2	32	34	52	54	20	3-Time estimate	0.3333
19	S	Yes	9	32	41	32	41	0	3-Time estimate	0.6667
20	T	no	6	34	40	54	60	20	3-Time estimate	0.6667
21	U	Yes	12	41	53	41	53	0	3-Time estimate	2.6667
22	V	no	17	41	58	47	64	6	3-Time estimate	1.6667
23	W	no	4	40	44	60	64	20	3-Time estimate	0.6667
24	X	Yes	11	53	64	53	64	0	3-Time estimate	1.3333
	Project	Completion	Time	=	64	weeks				
	Number of	Critical	Path(s)	=	1					

FIGURE 8–16a   The BIGEX project activity table input to QSB

---

*Business WinQSB, Version 1; Yih-Long Chang and Robert S. Sullivan, published by John Wiley, 1997.

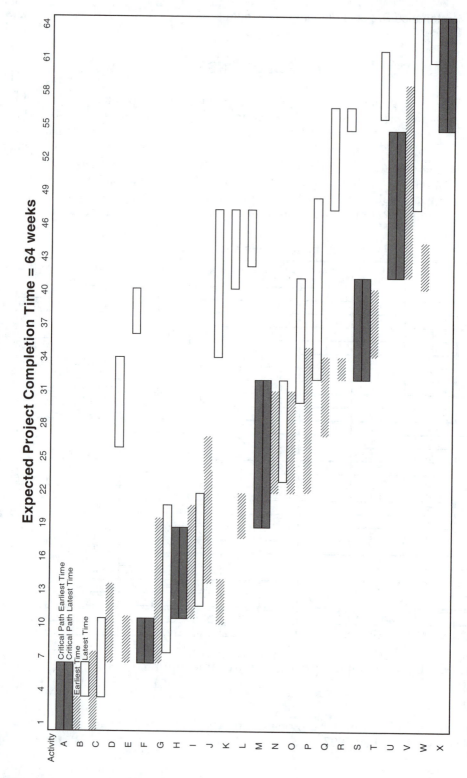

FIGURE 8–16b The BIGEX project Gantt chart QSB

Expected Project Completion Time = 64 weeks

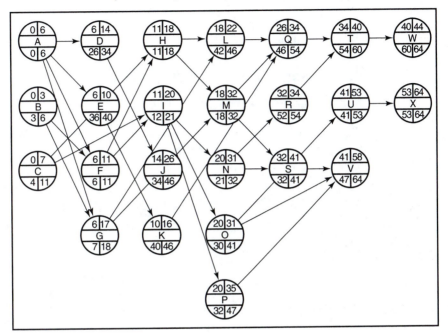

FIGURE 8–16c   An activity-on-node network diagram QSB output

Probability Analysis for BIGEX

05-26-2005 08:40:32	Critical Path	Completion Time Std. Dev.	Probability to Finish in 67 weeks
1	A–> F–> H–> M–> S–> U–> X	4.2817	75.8232%

FIGURE 8–16d   The QSB probability solution to Question 1

Probability Analysis for BIGEX

05-26-2005 08:41:24	Critical Path	Completion Time Std. Dev.	Probability to Finish in 69 weeks
1	A–> F–> H–> M–> S–> U–> X	4.2817	87.8556%

FIGURE 8–16e   The QSB probability solution to Question 2

QSB probability solutions are always for probability to complete *within* a given number of periods or less (<=). Thus, the computer solution to Question 2 (>69) is solved as $1 - .878556 = .121444$, or about 12%, as was our calculation.

8.5.1 Activity-on-Node Network Diagram Convention

Most computer software programs for PERT and CPM network analysis print out activity-on-node network diagram solutions. Refer back to Figure 8–16c in the previous section. That presentation shows the ES and EF times for each activity on the upper part of the lettered node circles, and LS, LF times on the lower part of the circles.

8.5.2 More Computer Printout Examples

Figure 8–17a is the WinQSB activity table input for the original construction of a manufacturing plant example from Section 8.4. The related network diagram printout is shown in Figure 8–17b, which is equivalent to our original activity-on-arrow solution diagram from Figure 8–11. In this diagram, the critical path is the A → C → E → F → I sequence of nodes. The WinQSB project analysis solution table is presented in Figure 8–17c.

Note that the activity-on-node diagram convention eliminates the need for dummy activities to show multiple activity precedents. It does, however, result in some crossover arrows between the nodes to indicate precedence relationships (refer again to Figure 8–16c for the complex example from the previous section). The crossover arrows should not hinder the analyst's correct interpretation of the diagram.

Project Problem

Activity Number	Activity Name	Immediate Predecessor (list number/name, separated by ',')	Normal Time
1	A		12
2	B		10
3	C	A,B	3
4	D	C	23
5	E	C	17
6	F	E	8
7	G	C	12
8	H	G	8
9	I	D,F,H	5

FIGURE 8–17a   The WinQSB input data for plant construction example

Project Completion Time = 45 weeks

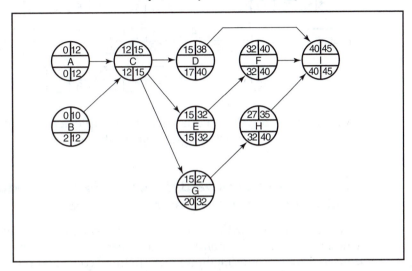

FIGURE 8–17b The WinQSB activity-on-node network diagram for the example project

Activity Analysis for Project Problem

06-01-2005 11:26:15	Activity Name	On Critical Path	Activity Time	Earliest Start	Earliest Finish	Latest Start	Latest Finish	Slack (LS-ES)
1	A	Yes	12	0	12	0	12	0
2	B	no	10	0	10	2	12	2
3	C	Yes	3	12	15	12	15	0
4	D	no	23	15	38	17	40	2
5	E	Yes	17	15	32	15	32	0
6	F	Yes	8	32	40	32	40	0
7	G	no	12	15	27	20	32	5
8	H	no	8	27	35	32	40	5
9	I	Yes	5	40	45	40	45	0
	Project	Completion	Time	=	45	weeks		
	Number of	Critical	Path(s)	=	1			

FIGURE 8–17c The WinQSB analysis solution table for the example project

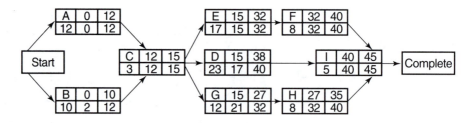

FIGURE 8–18   Plant construction example activity-on-node network diagram calculations

8.5.3   Manual Forward/Backward Pass Procedures on Activity-on-Node Diagrams

For manual forward and backward pass calculations using activity-on-node networks, it is helpful to show each activity's time estimate in its node. Square nodes facilitate these manual computations in some widely used texts. Interested readers are referred to other sources on the subject.

Figure 8–18 is our original manufacturing plant construction example network diagram in a format shown in the Anderson/Sweeney/Williams management science text.[*] Activity time is shown below the letter identifying each activity. The upper two values are the ES and EF times; the lower values are the LS and LF times.

Starting with the activity letter designator and the estimated time in each node square in the diagram, it is easy to work the forward and backward pass procedure, adding ES, EF, LS, and LF values in each node. This technique is the same as when annotating activity *arrows* with the calculated values in the activity-on-arrows diagrams we examined earlier in the chapter.

Questions

1. Why are projects generally not defined by repetitive tasks?
2. Identify the characteristics of maintenance projects.
3. Explain why one aspect of the definition of a project is that it could be delayed to a future date.

---

[*]Anderson, David R., Dennis J. Sweeney, and Thomas A. Williams. *An Introduction to Management Science: Quantitative Approaches to Decision Making,* 2d ed. St. Paul, MN: West Publishing Company, 1979.

4. Identify three categories of questions that should be answered to guarantee project success.

5. List the nine steps that are key to project success.

6. Define the term *precedence relationship*.

7. Think of a small project you have worked on. Identify the work breakdown structure and identify any precedence relationships.

8. What is meant by the term *critical path*?

9. Explain the difference between internal and external boundaries.

10. Explain the difference between the two types of maintenance project scheduling methods.

11. Identify at least three parts to a maintenance management program.

12. What is the distinction between PERT and CPM as network tools?

13. What does it mean to *crash* an activity in project scheduling?

CPM Crash Analyses

These problems provide a good review of the CPM and PERT analysis techniques. Ensure that you solve them correctly by checking the given solutions. Entering the exercise activity tables into *network models* software also would be worth the effort. That would again confirm the chapter's point that, given your understanding of network model computational procedures, computer software is the efficient way to avoid tedious, repetitive network calculations.

14. Given the following project activity table below, draw the descriptive network diagram using either the activity-on-arrow or activity-on-node convention.

Activity*	Immediate predecessor(s)	Times (weeks) Normal	Crash	$Costs Normal	Crash
A	—	9	7	8,500	10,500
B	—	8	6	4,000	6,400
C	A	6	5	7,000	8,500
D	A	4	3	2,200	2,800
E	B	5	4	2,500	3,300
F	C	5	4	5,000	5,800
G	D, E	6	4	6,500	8,100

*Activities F and G complete the project.

15. Refer back to Question 14 to perform the following analyses:
   a. Perform a normal activity time network analysis (forward and backward passes) to determine project time, critical path of activities, and normal project cost.
   b. Shorten the project to two weeks less than the normal time in the most cost-effective way. Which activity is reduced one week on the first step? Which activity is reduced one week on the second step? After the two-week reduction, what is the project time, critical path(s), and project cost?

16. Following is a more complex CPM project activity table.
   a. Draw the related network diagram.
   b. Perform a normal-time network analysis for project time, critical path, and cost.
   c. Perform an all-activities-at-full-crash-time analysis to calculate project time, critical path(s), and cost.
   d. Starting with the normal-time analysis of part (b), do iterative steps of one-week reduction on each step to determine how the minimum project time of part (c) can be achieved without crashing *all* project activities. Show the resulting times for each activity, the critical paths, and project cost.
   e. How much cost savings does the part (d) analysis represent over the full crash concept of part (b)?

Activity	Immediate predecessor(s)	Times (days)		$Costs (dollars)	
		Normal	Crash	Normal	Crash
A	—	10	8	$2,000	$2,400
B	—	8	7	500	750
C	A	13	10	800	1,700
D	A	11	11	750	750
E	B	12	9	1,500	1,800
F	B	15	12	650	875
G	C	15	12	1,600	1,900
H	C	7	6	400	800
I	D, E	8	5	600	1,050
J	F, H, I	13	12	800	1,050

*Activities G and J complete the project.

PERT Analyses

17. Suppose an analysis of a large PERT-managed project disclosed a long, multiactivity critical path along which the mean project time was 118 weeks. Project *variance* along the critical path is 144 weeks.

a. There is some concern that the project might require more than 136 weeks, requiring resources that they had hoped to free up for use on a new project. What is the probability that the project *will* take more than 136 weeks? (Answer to nearest percent.)

b. The activity supervisors will earn bonuses if the project is completed in less than 114 weeks. What is the probability of that bonus event? (Answer to nearest percent.)

18. Following is the activity schedule for a project to be planned and managed by PERT.

a. Draw the network diagram.

b. Complete a forward and backward pass to determine the project average time and critical path sequence.

c. Calculate the project time standard deviation value to two decimal places.

d. The project manager would be concerned if there were greater than 5% probability that the project completion will exceed 25 months. Should she be worried? Clearly show supporting probability computations.

Activity*	Immediate predecessor(s)	Times estimates (weeks)		
		a	m	b
A	—	5	7	9
B	—	5	7	15
C	B, —	0.5	2	3.5
D	A, B, —	2	6.5	8
E	A, B	6	9	12
F	B	2	3	4
G	C, D	4.5	6	7.5
H	E	1	4	7
I	F	7	9	11

*Activities G, H, and I complete the project.

19. Let's conclude the CPM/PERT review problems with a more realistically sized project, using the following activity schedule. Perform the necessary average activity and project time computations, with the related variances. Do the necessary network analyses. Then compute the probabilities of these project time values considered important for planning this project:

a. The probability that the project could be done in 42 weeks or less.

b. The probability that the project will exceed 48 weeks.

Activity*	Immediate predecessor(s)	Times estimates (weeks)		
		a	m	b
A	—	4	6	8
B	—	1	4.5	5
C	—	4	5	6
D	—	1.5	3	4.5
E	A	3.5	4	4.5
F	A	4	4.5	8
G	B	6	7	8
H	D	2	4	6
I	C, H	6	7	14
J	D	8	11	14
K	E	13	15	17
L	E	9	12.5	19
M	E	2	5	8
N	F, G, I, M	8	8	8
O	C, H	9	18	21
P	J	6	7	8
Q	J	12	17.5	20
R	L, N, O, P	4.5	6	7.5
S	L, N, O, P	10.5	12	13.5
T	K, R	11	14	17

*Activities Q, S, and T complete the project.

9

COMPUTERIZED MAINTENANCE MANAGEMENT SYSTEMS

Overview

Objectives

At the completion of the chapter, students should be able to

- Explain CMMS and its functions.
- Understand the role of CMMS in modern maintenance management.

9.1   INTRODUCTION

Significant advances in computer hardware and software development have affected most areas of business and industry, and the arena of maintenance planning and management is no exception. The use of computerized maintenance management systems, which are commonly referred to as CMMS, is no longer a luxury or a frivolous business overhead; in many cases, it is a requirement. Enterprises that want to attain ISO or QS certification will discover that application of CMMS is a fundamental requirement to successfully obtain and maintain such certifications.

A variety of software packages are available, and many have been around for a number of years. Some of the older programs have disappeared as more state-of-the-art packages have been introduced. The advent of faster and more powerful computers, the development of user-friendly, menu-driven software packages, and the specific tracking and reporting requirements set forth by customers and certification bodies such as ISO and QS have made the use of the technology much more appealing and, therefore, more commonplace.

Today, CMMS are used for all aspects of maintenance planning, management, and control. Activities including (but not limited to) maintaining equipment history, scheduling preventive and predictive maintenance activities, tracking personnel skills and assigning maintenance tasks based on these skills, streamlining spare parts inventory management, and producing a vast array of management reports are routinely performed by computers.

Other maintenance-related activities such as planning alternate scheduling scenarios, reliability studies, and root cause analysis can benefit from sophisticated and advanced software programs that make use of computer simulation and modeling. These programs allow the analyst to review several alternatives and scenarios that are simulated using animation and real-life characteristics. A major advantage of computer simulation is the nondestructive nature of testing. Another advantage of computer simulation is the ability to accelerate the time factor of the process.

There are a number of user-friendly, advanced CMMS packages available. Whereas some may offer special features unique to that package, most offer the same basic features. Some packages may contain advanced modules structured around the program that allow the user a certain degree of flexibility in customizing the package and selecting only those functions deemed necessary. Of course, prices vary depending on the level of software's sophistication and ability.

This chapter by no means intends to survey available computerized maintenance management systems. Rather, the purpose is to take a cursory look at some of the functions provided by CMMS software and to present a brief overview of some of the more interesting features that a company may expect from the program.

We will use the CMMS package MP2, developed and marketed by Datastream Systems, Inc., as our example. According to the 1995 Booth Research Study, Datastream, with more than 41,000 installations worldwide and in excess of 50% of the market share among the top 10 CMMS vendors, was selected as the world leader in maintenance solutions. Some of the features of MP2 are discussed briefly to familiarize the reader with the standard CMMS functions.

The case study presented at the end of the chapter introduces MAP-CON, which is another well-respected CMMS.

9.2   BASIC FUNCTIONS

9.2.1   Labor Tracking

For scheduling, and in order to match various maintenance activities with the proper craft codes, it is necessary to maintain an accurate and complete inventory of personnel skills. A CMMS package must have the ability to maintain records of craft skills, training and educational information such as course description hours and credit units earned, and other relevant data as part of the employee records for all individuals involved in the maintenance process. Figure 9–1 presents a data screen displaying a

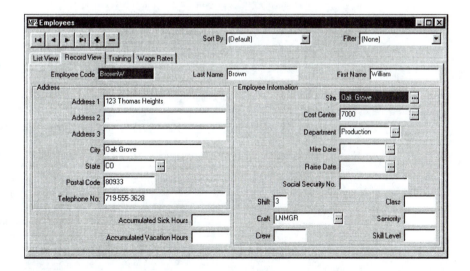

FIGURE 9–1   A maintenance employee record

(Source: Copyright © Datastream Systems, Inc. All rights reserved. Used by permission.)

maintenance employee's personnel record. The system also allows multiple wage rates to be entered for different kinds of work, and the wages for each task performed can be charged to the appropriate cost centers.

9.2.2 Vendor and Manufacturer Information

An important part of any maintenance organization's resources is information about all vendors and contractors that have sold equipment or provided services to the company. CMMS packages should have the ability to maintain records on the manufacturers of the equipment located at the site. Records should contain basic data such as each vendor's address and contact person. In addition, other pertinent information such as a list of all items and equipment supplied by each vendor, ordering methods, terms and contractual agreements, and relevant comments also should be recorded. Samples of vendor records are displayed in Figure 9–2.

9.2.3 Inventory Management

An effective spare parts inventory management system is part of every good maintenance management program. The CMMS should keep track

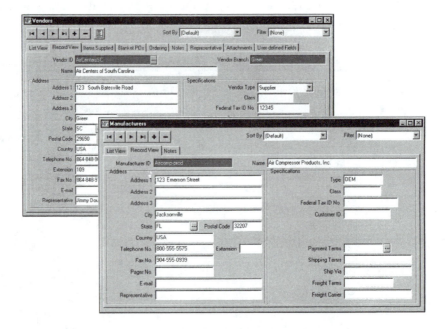

FIGURE 9–2   Vendor and manufacturer information

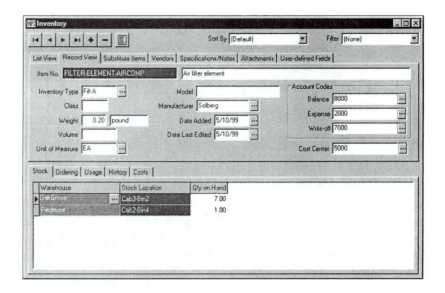

FIGURE 9–3   Spare parts inventory record
Copyright © Datastream Systems, Inc. All rights reserved. Used by permission.

of all facilities maintenance parts, recording part numbers, stock location, quantities on hand, and unit cost. The system also should manage
the reordering process to avoid downtime caused by stockout and
reduce the carrying costs of excessive inventory. Figures 9–3 and 9–4
display some of the features of parts inventory records.

Additional features that are useful to include in the inventory management system are a list of parts that may be substituted for the inventory item; cross references and links to all the vendor records that supply
the part; reordering information such as EOQ, inventory tracking, and
ABC classification and usage history; and all engineering and manufacturing specifications for each part.

9.2.4   Equipment Records

The features provided in CMMS support the maintenance activities of
the equipment. Therefore, equipment records are the heart of the CMMS
system. *Equipment* refers to any object on which maintenance is tracked.
Equipment may be related to production, such as presses and so on, or
can be nonproduction equipment such as material handling or other
peripheral systems. Equipment also may refer to any of the subassem-
blies or components of a larger system.

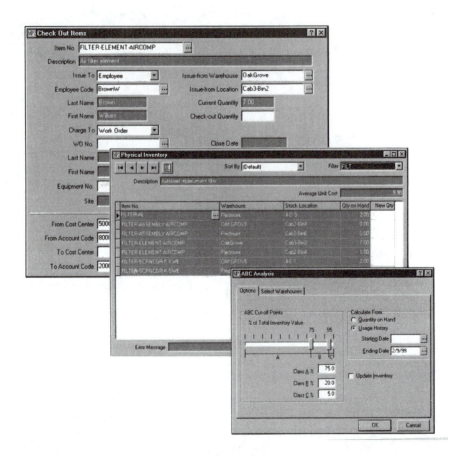

FIGURE 9–4   Additional features of the spare parts inventory record
Copyright © Datastream Systems, Inc. All rights reserved. Used by permission.

Equipment data such as number, type, model and serial numbers, manufacturer, and purchase date are the essential fields of information in equipment records. Equipment site and location also must be specified. Cost data such as the original purchase price and the associated cost center also are maintained. Spare parts data and hierarchical relationships between the equipment and its components are part of the equipment record as well.

Equipment histories, warranty data, and service contract records are maintained in the records for each piece of equipment, and the CMMS specifies if the service contract or warranty covers the equipment and needed repairs. Performing failure analysis should be an integral part of CMMS. Data about equipment failure for each equipment type, including the reason for the breakdown and how the problem was

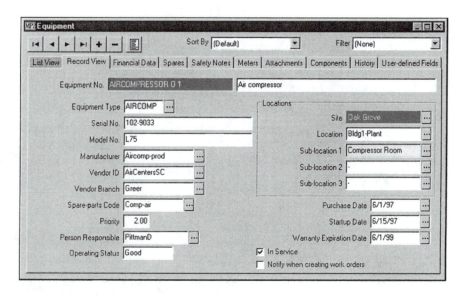

FIGURE 9–5 Equipment data screen

Copyright © Datastream Systems, Inc. All rights reserved. Used by permission.

solved, can be used to establish and analyze breakdown trends and aid with the troubleshooting process. Safety information about the equipment is included as part of the equipment record. Additionally, all related OSHA (Occupational Safety and Health Administration) regulations that affect the organization are maintained and included in the work orders to remind personnel about the regulatory requirements. Sample equipment screens are shown in Figures 9–5 and 9–6.

9.2.5 Scheduling

Automatic scheduling of maintenance activities is an important part of any CMMS package. Scheduling should be based on the equipment's need, availability of qualified personnel, and the production schedule. When scheduling maintenance activities, CMMS allows the user to specify workdays and exceptions to those days for the site and for each employee for such reasons as vacation, sick days, holidays, and so on.

Current work orders and tasks can be viewed by day, week, or month. Additional tasks can be assigned for underutilized days, or the load may be reduced if the schedule becomes overloaded in a given period. Scheduling input data are presented in Figure 9–7.

CMMS can accomplish the scheduling of maintenance activities according to site, employee, or work order scheduling. Site scheduling

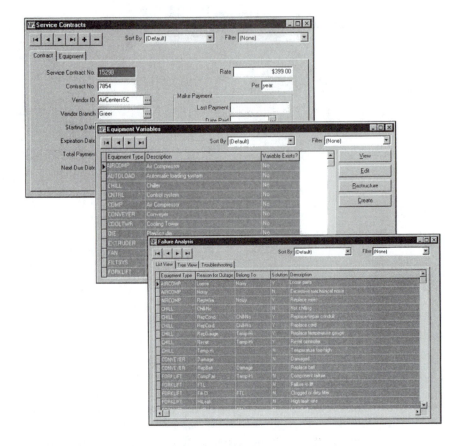

FIGURE 9–6   Additional equipment data screen
Copyright © Datastream Systems, Inc. All rights reserved. Used by permission.

avoids generating work orders for days that the plant does not operate. Employee scheduling is based on available employee hours. The system keeps track of each employee's extra or missed work hours. All work orders for a certain craft or a specific employee, as well as all due tasks or work orders for a given day, week, or month, can be viewed. Backlogs and unscheduled work orders are also available for perusal.

9.2.6   Predictive Maintenance

Statistical predictive maintenance features of CMMS can identify equipment readings outside the control limits and alert you to schedule the required maintenance procedures before the equipment fails. These control

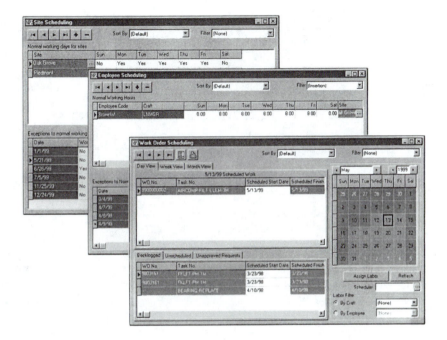

FIGURE 9–7 Scheduling features
Copyright © Datastream Systems, Inc. All rights reserved. Used by permission.

limits may be set based on either the manufacturer's specifications or the equipment's historical performance. Periodic graphs and reports show the readings and the trend of a specific metric, for instance, the equipment temperature or vibration readings. These readings are compared with the minimum and the maximum levels that are set according to the vendor's specifications in order to detect any out-of-control conditions.

Potential equipment failure can be determined based on comparison with the manufacturer's specification, average readings, standard deviation, or data trends, and may be based on multiple metrics. Work orders for preventive maintenance actions can be generated automatically once these out-of-control conditions are detected. An example of a statistical predictive maintenance screen is displayed in Figure 9–8.

9.2.7 Work Orders

Work orders are generated automatically for scheduled preventive maintenance actions, or they may be created manually for corrective and emergency situations. Sample screens displaying work order data are

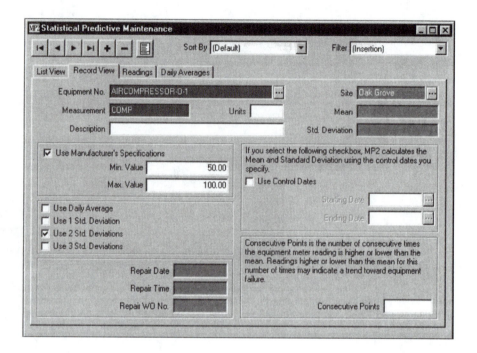

FIGURE 9–8   Statistical predictive maintenance input data
Copyright © Datastream Systems, Inc. All rights reserved. Used by permission.

presented in Figure 9–9. When the task is completed the, repair and labor data can be entered to update historical information. The CMMS transfers work order history and updates all relevant equipment and task records.

Work orders may be generated for single or multiple equipment tasks, for "ready" or "hold" tasks, or for specific tasks, locations, or equipment. CMMS also is capable of attaching relevant documents such as drawings or other specific instructions to a work order.

Work also can be *requested*. The system allows for either call-in or on-site work requests. Maintenance request information is entered into the system as it occurs. On-site work request capability allows the employees to enter their own maintenance requests.

9.2.8   Purchase Orders

As mentioned previously, spare parts inventory management is an important feature of CMMS. In addition to tracking the spare parts inventory and updating records according to receipts, returns issues, and so on, purchase orders for parts can be issued from requisitions or they may

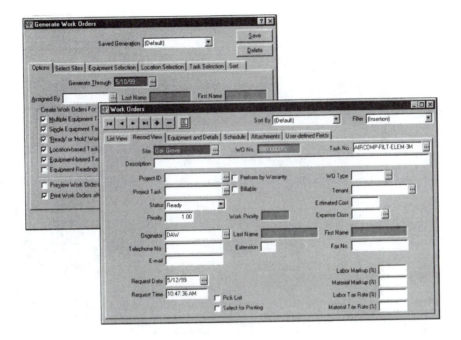

FIGURE 9–9   Sample work order data
Copyright © Datastream Systems, Inc. All rights reserved. Used by permission.

be generated manually for noninventory items. A complete history of all purchase order transitions is maintained by the system.

9.2.9   Security

CMMS should provide a common database across the facility so that all information is entered only once and current data is always available when needed and on a *need-to-know* basis. Security features allow for setting security at various levels so that only users who are authorized may access specific information in the database. Figure 9–10 shows various security input data.

9.2.10   Reports

CMMS is capable of generating an array of reports as well as analytical and graphical data. More than 160 different types of reports in more than 15 distinct areas, such as statistical preventive maintenance, work requests and orders, scheduling, inventory, and labor management summaries are available. These reports provide an invaluable tool for managing the resources of the enterprise.

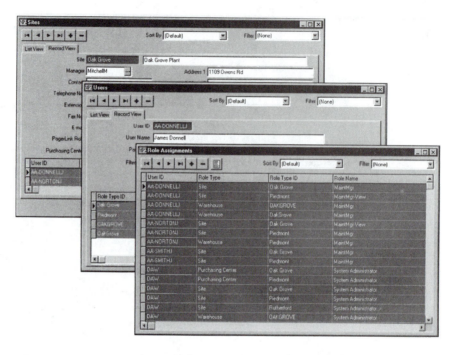

FIGURE 9–10   Security setup data screens
Copyright © Datastream Systems, Inc. All rights reserved. Used by permission.

CMMS AND ISO/QS CERTIFICATION

Attaining and maintaining ISO or QS certification requires strict adherence to a system of standards established by the respective international committee. In each case, the governing precepts or the "golden rules" of certification are

- Say what you do: Document standard operating procedures (SOP).
- Do what you say: Adhere to these SOPs.

- Document the two previous statements.

Several years ago, the ISO standards were revised to require the use of computerized maintenance management systems (CMMS) in place of the archaic and the error-prone manual method.

WCI Steel, Inc., in Warren, Ohio, has been quite successful in attaining and maintaining their ISO Certification with the aid of MAPCON, a powerful CMMS. This case study briefly examines WCI Steel, Inc.'s success with MAPCON.

Say What You Do: Documentation of SOPs

MAPCON is an excellent and easy-to-use tool for creating and editing the control processes for both maintenance and operations. MAPCON security provides the necessary control features for creating and editing SOPs. The most up-to-date changes then are made available for on-screen viewing throughout the plant. MAPCON eliminates the need for developing procedures for handling and removing outdated information, which is inherent in a manual system. The possibility of access to outdated information, a definite ISO noncompliance issue, therefore is eliminated. Furthermore, since not all procedures are ISO-related procedures, MAPCON allows SOPs to be flagged as either ISO or non-ISO procedures for subsequent sorting and reporting.

In order to assure that maintenance functions are performed safely, safety procedures, some required by law, are defined for nearly every maintenance function. Safety maintenance procedures outline the steps for the task and are required for many ISO/QS procedures for equipment maintenance. As illustrated in Figure 9–11, MAPCON

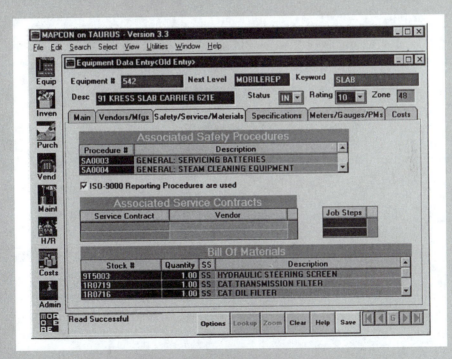

FIGURE 9–11 MAPCON's equipment data entry screen
(Source: Courtesy of MAPCON Technologies, Inc.)

(continued on next page)

attaches safety procedures to equipment records as well as to PM routines. When work orders for a piece of equipment are generated, safety procedures are attached automatically.

ISO/QS standards require that equipment with "a direct impact on the quality of the product" be identified. MAPCON enables the user to identify each piece of equipment as either ISO- or non-ISO critical.

Along with ISO-equipment criticality, ISO/QS certification requires the identification of employees who have the required skills to maintain ISO-critical equipment. MAPCON uses a series of craft codes that are defined in the software and can access employee records to match the required skills with those of an employee. As employees participate in various training programs, the time-card system automatically updates employees' records to show newly acquired skills.

The spare parts inventory and management features in MAPCON allow tracking of all essential inventory and items with long lead times. This ensures that the maintenance organization has the tools to ascertain that the right parts are available when needed, so that essential maintenance on critical equipment can be performed in a timely manner.

Do What You Say: Adherence to SOPs

The most important feature of any CMMS is its ability to monitor and manage the preventive maintenance (PM) activities of the enterprise. All the features of the software, from the equipment data entry system to spare parts inventory tracking to monitoring equipment criticality and employees' skills, only serve to support the PM features of MAPCON. PM procedures detail the required steps specified by each equipment manufacturer to maximize equipment uptime. Based on equipment usage and operating conditions, manufacturers' minimum standards can be adjusted as needed. Once established, MAPCON tracks and sets all due dates for various PM steps. PM work orders are generated and assigned to the appropriate personnel according to their skills determined by the required craft codes. As downtime decreases, the overall costs decrease and quality and productivity increase.

Documentation

MAPCON provides a wide range of report types to help the user document all aspects of the process, which is necessary to satisfy adherence and compliance with the ISO/QS standards. These reports can be generated to substantiate that the company "says what it does" and "does what it says." Figure 9–12 shows the MAPCON reports menu used by WCI Steel, Inc.

Source

Johnson, E., CMMS manager, WCI Steel, Inc. *ISO/QS 9000 Documentation via MAPCON* (Internal report).

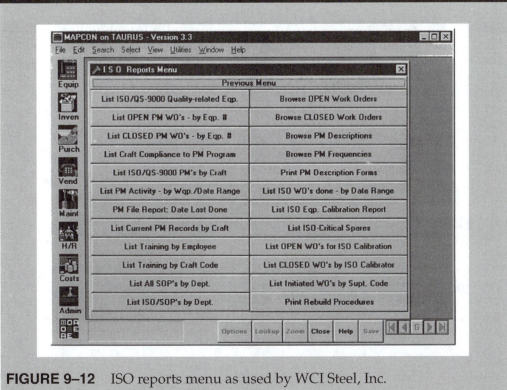

FIGURE 9–12   ISO reports menu as used by WCI Steel, Inc.
(Source: Courtesy of MAPCON Technologies, Inc.)

QUESTIONS

1. What is the role of CMMS in modern maintenance management?
2. How can CMMS help with maintenance management control functions such as inventory management and tracking?
3. How does CMMS help with labor tracking?
4. How does CMMS help with tracking and maintaining equipment history?
5. What is the role and function of CMMS in complying with OSHA and other regulatory requirements?
6. What role can CMMS play in devising and adhering to a workable maintenance schedule?

7. How can CMMS help create a preventive maintenance program?

8. What is the statistical ability of most CMMS packages?

9. What role can CMMS play in obtaining and maintaining ISO and QS certifications?

SOLUTIONS AND ANSWERS TO PROBLEMS AND QUESTIONS

In this section, detailed solutions and answers to selected questions from the end-of-chapter exercises are provided. Use the solutions as an additional aid in the learning process, to enhance learning and to develop mastery in the subject matter. It is therefore strongly suggested that you make a serious effort to answer the questions without referring to the answers that are provided here. When you need guidance or simply want to check your answers, then these solutions can provide a reinforcement in the learning process.

Chapter 1—Introduction

1. The increased complexity of modern manufacturing systems highlights the need for a highly skilled maintenance staff and specialized planning, training, and development of programs for maintenance activities and maintenance personnel.

2. **a.** Maintenance: All activities necessary to keep a system and all its components in working order.
   b. Failure: Any deviation or change in a product or system from a satisfactory working condition to a condition that is below the acceptable or set operating standards for the system.

3. Corrective maintenance: The required repair work after the equipment failure has occurred.

   Preventive maintenance: Takes steps to prevent and fix problems before failures occur.

   Predictive maintenance: Predicts possible failures using statistical tools and various instruments and tests, such as vibration analysis, chemical analyses of lubricants, thermography, optical tools, and audio gauges.

4. Any unnecessary expenditures that can be eliminated by implementing a sound maintenance program will result in reduction of expense, hence an increase in profit.

5. The optimum prevention level is the point at which the total costs are at the minimum. (See Figure 1–1.)

6. Primary objectives of maintenance include:
   - Maintaining existing equipment
       Timely and appropriate response to equipment failure, reduction of downtime, increase equipment availability
   - Equipment inspection, cleaning, and lubrication
       Develop program for operators to do routine tasks to detect problems before they occur, develop schedule for regular and routine cleaning and lubrication
   - Equipment modification, alteration, and installation
       Nonroutine activities, so can be scheduled during slack periods to increase efficiency of maintenance personnel
   - Utility generation, distribution, and management
       Maintenance and efficient operation of steam, electricity, etc.
   - Maintaining existing buildings and grounds
       Building repairs, painting, etc.
   - Building modification and alteration
       Plant expansions, process changes

7. Secondary functions of maintenance include
   - Plant protection and security
   - Salvage of obsolete equipment and waste disposal
   - Pollution and noise control
   - ADA, EPA, OSHA, and other regulatory compliance

8. When scheduling maintenance activities, the scheduler must consider the priority of various tasks, the amount of time required for completion, the type of tools, equipment, and labor needed, and the size and skill of maintenance crew needed.

9. Maintenance backlog generally can be defined as how much has to be done. Keeping the backlog at or near zero level is not recommended because the maintenance personnel will be idle when no repairs need to be made. An average backlog of two to three weeks is recommended.

10. Crew size = Scheduled labor hours / (Backlog × Hours per week)
    Crew size = 1500 hours / (4 weeks × 35 hours per week)
    Crew size = 10.71 or 11 people

11. Backlog = scheduled hours per week / (Crew size × Hours per week)
    Backlog = 1200 hours / (15 people × 40 hrs per week)
    Backlog = 2 weeks

12. CMMS: Computerized maintenance management systems
    - Used for maintenance management planning and control
    - Roles include staffing, effective planning and scheduling of events, ability to track and control backlog, creating and tracking equipment history and work orders

13. TPM: Total productive maintenance, the continuous and overall improvement in equipment effectiveness through the active involvement and participation of all employees.

14. The operator is the key participant in the TPM environment. TPM utilizes all available resources including operators to improve and maintain equipment. This empowers the operators and allows them to take ownership of the equipment and assume responsibility for basic and routine maintenance activities.

15. TPM programs strive to reduce and eliminate six significant obstacles to achieve full equipment effectiveness:
    1. Equipment failure: Reduce equipment breakdown.
    2. Setup and routine adjustments: Increase equipment utilization.
    3. Idling and stoppage: Trained operator can detect and remedy routine causes of abnormalities and slowdowns.
    4. Reduce speed: Trained operators will be able to detect the deviations between expected and actual speeds.
    5. Defects: Scrap and defects due to out-of-control conditions can be recognized and corrected.
    6. Startup problems: Recognize and correct problems associated with achieving a stable process.

16. Centralized
    Advantages
    - Better utilization of human and equipment resources
    - Overall backlog can be controlled
    - Better control of inventory of special equipment and material
    Disadvantages
    - Time lost in traveling among various locations
    - Scheduling can be cumbersome
    - Less supervision of maintenance crew

    Decentralized
    Advantages
    - Various departments have their own maintenance departments
    - Faster response to maintenance needs
    - Better supervision of crew
    Disadvantage
    - Duplication of personnel, special maintenance equipment, and material, which increases inefficiency and cost

Chapter 2—Statistical Applications

1. **a.**

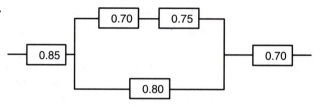

$R = (0.70)(0.75) = 0.525$ 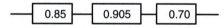

$R = 1 - (1 - 0.525)(1 - 0.80) = 0.905$

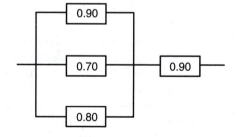

$R = (0.85)(0.905)(0.70) = 0.538475 = \mathbf{53.8\%}$

b.

$R = 1 - (1 - 0.90)(1 - 0.70)(1 - 0.80) = 0.994$

$R = (0.994)(0.90) = 0.8946 = \textbf{89.46\%}$

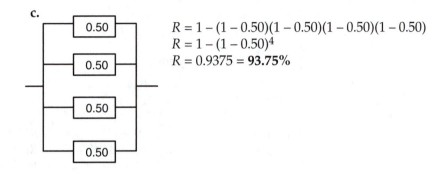

c.

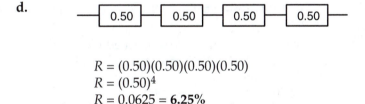

$R = 1 - (1 - 0.50)(1 - 0.50)(1 - 0.50)(1 - 0.50)$
$R = 1 - (1 - 0.50)^4$
$R = 0.9375 = \textbf{93.75\%}$

d.

$R = (0.50)(0.50)(0.50)(0.50)$
$R = (0.50)^4$
$R = 0.0625 = \textbf{6.25\%}$

2. a. λ = number of failures / number of hrs tested
$4/(20 + 24 + 45 + 65 + (2*90)) = 4/334 = \textbf{0.01198}$
b. MTBF = $1/\lambda = 1/0.01198 = \textbf{83.47 hrs}$
c. Availability = MTBF/ (MTBF + MDT) = $83.47/(83.47 + 30) =$
$0.7356 = \textbf{73.56\%}$
d. Reliability @ $t = 1$
  $R(1) = e^{-\lambda t}$
  $R(1) = e^{-(0.01198)(1)}$
  $R(1) = 0.98809 = \textbf{98.81\%}$
e. Reliability @ $t = $ MTBF
  $R(\text{MTBF}) = e^{-\lambda t}$
  $R(83.47) = e^{-(0.01198)(83.47)}$
  $R(83.47) = e^{-(0.9999706)}$
  $R(83.47) = 0.36789 = \textbf{36.79\%}$
3. a. λ = number of failures / number of hours tested
  $= 5/(45 + 55 + 80 + 105 + 110 + (3*120)) = 5/755 = \textbf{0.0066225}$
b. MTBF = $1/\lambda = 1/0.0066225 = \textbf{151 hours}$

    c. Availability = MTBF / (MTBF + MDT) = 151 / (151 + 60) = 151/211 = 0.7156 or **71.56%**

    d. $R(1) = e^{-\lambda t}$
      $R(1) = e^{(-0.0066225)(1)}$
      $R(1) = 0.9934$ or **99.34%**

    e. $R(\text{MTBF}) = e^{-\lambda t}$
      $R(151) = e^{-(0.0066225)(151)}$
      $R(151) = e^{-0.9999975}$
      $R(151) = 0.3679$ or **36.79%**

4. **a.** $\lambda = 1/\text{MTBF} = 1/250 = 0.004$

    b.

$R(150) = e^{-\lambda t}$	$R(250) = e^{-\lambda t}$	$R(350) = e^{-\lambda t}$
$R(150) = e^{-(0.004)(150)}$	$R(250) = e^{-(0.004)(250)}$	$R(350) = e^{-(0.004)(350)}$
$R(150) = e^{-0.6}$	$R(250) = e^{-1}$	$R(350) = e^{-1.4}$
$R(150) = 0.5488$	$R(250) = 0.3679$	$R(350) = 0.2466$
$R(150) = \mathbf{54.88\%}$	$R(250) = \mathbf{36.79\%}$	$R(350) = \mathbf{24.66\%}$

5. 300 days × 8 hours/day = 2400 hours total
2400 hours / 200 hours = 12 failures per year
12 failures × $5000 per failure = $60000 per year (without mainte-nance)

2400 hrs / 120 hrs = 20 scheduled PM
20 scheduled PM × $450 per PM = $9000
20 scheduled PM × 0.35 = 7 failures
7 failures × $5000 = $35000
total cost of PM = $9000 + $35000 = $44000

$44,000 is still less than $60,000, therefore this PM schedule is recommended.

[handwritten left margin]
if $(0.35 \times \lambda \times 2400)$
=> 4.2 failures
∴ Failure cost = $21,000
    not
    $35,000

6. $\lambda = 0.025$

    a. $R(\text{MTBF}) = e^{-\lambda t}$ MTBF $= 1/\lambda = 1 / 0.025 = 40$
      $R(40) = e^{-(0.025)(40)}$
      $R(40) = e^{-1}$
      $R(40) = 0.3679$ or **36.79%**

    b. PM if $R = 0.75 = ?$
      $R = e^{-\lambda t}$
      $0.75 = e^{-0.025t}$
      $\ln 0.75 = -0.025t$
      $-0.257682 = -0.025t$
      $t = \mathbf{11.5 \ hours}$

7. $\lambda = 0.002$
PM if $R = 0.80 = ?$
$R = e^{-\lambda t}$
$0.80 = e^{-0.002t}$
$\ln 0.80 = -0.002t$

$$-0.22314 = -0.002t$$
$$t = \textbf{111.57 hours}$$

Chapter 3—Preventive Maintenance

1. Preventive maintenance: A series of predefined and scheduled maintenance activities that are designed to reduce equipment breakdown, increase equipment reliability, and improve productivity.

2. Economic considerations are important because an increase in the level of PM equals an increase in the total cost level.

3. Ideally, all maintenance activities should be preventive.

4. Empowerment signifies a sense of ownership of the operators' equipment. It makes the operator feel like they have control over their equipment. The operator can play a key role in TPM by doing basic routine activities including cleaning, lubricating, and inspecting.

5. Basic PM activities include cleaning, lubricating, and inspecting. They are important because a problem can be detected before it turns into a major catastrophe.

6. Equipment history is important because it aids in determining equipment reliability, MTBF, and equipment availability. It also can point out failure trends and causes of failures.

7. A system of criticality is a system that classifies equipment based on the degree to which the equipment and its failures are considered to be critical.

   Criticality Code I: No backup, will cause harm, stop production

   Criticality Code II: Breakdown undesirable but not drastic, slowdown in production, might cause harm

   Criticality Code III: Doesn't seriously affect normal operation, won't cause harm

8. A checklist is important for PM because it allows for the same things to be done to the equipment every time PM is performed.

9. Planning for maintenance should begin at the design stage. Through proper design, the need for and frequency of some maintenance activities can be drastically reduced and many functions can be simplified.

10. "Design for maintainability" means that the product should be designed so that maintenance functions are easy to perform. This is because the more difficult the maintenance function, the more likely it will not be done, or will not be done correctly.

11. Open-ended discussion question—answers will vary.

12. Expected breakdowns = Σ [(#failures)(#months occurred) / total months]
    Expected breakdowns = [(0*4) + (3*6) + (5*6) + (8*5) + (9*3)] / 24
    Expected breakdowns = 115 / 24 = 4.79 per month

    4.79 breakdowns × $1050 per breakdown = **$5029.50**

Chapter 4—Predictive Maintenance

1. Predictive maintenance applies to various technologies and analytical tools used to measure and monitor various system and component operating characteristics and to compare these data with established and known standards and specifications in order to predict system or component failures, whereas corrective maintenance is applied after the failure, and preventive maintenance uses precautionary measures to avert possible problems.

2. For an industrial organization to be considered world class, one-third of all maintenance activities on average need to be predictive in nature.

3. The quantitative nature of predictive maintenance means that data are collected, analyzed, charted, and interpreted, and vital decisions are based on these data. This differs from PM because PM is scheduled maintenance to prevent failures, whereas PDM uses data to predict when a failure will occur.

4. Some of the methods of PDM are vibration analysis, chemical analysis, thermography, ultrasound techniques, radiography, radioscopy, and magnetic particle testing.

5. Vibration can significantly reduce the useful life of a product because excessive vibration can cause rapid equipment deterioration.

6. Some of the factors that determine the acceptable levels of vibration for a piece of equipment are size, stiffness, and weight of the equipment, the rigidity of the base on which it is mounted, and the surrounding equipment.

7. **a.** 0.28–0.71 mm/s
   b. 4.50–7.10 mm/s
   c. 4.50–7.10 mm/s

8. Tribology is the science and technology of friction, lubrication, and wear. It significant in PDM because understanding and controlling friction aid in material and energy conservation.

9. An oil analysis can help in PDM because it can inform operators of potential problems or failures that could occur as a result of contamination of lubricants.

10. The discovery of silicon in oil indicates that there is airborne dirt in the oil (see Table 4–4).

11. The purpose of thermography is to study various temperatures to detect potential problems in equipment using different methods and tools. In PDM, thermography can be used to locate "hot spots" and their possible sources.

12. Some applications of ultrasound in PDM include using a transducer to emit high frequency ultrasonic waves that are directed toward an object or are used to flood a shell or a part cavity.

13. The pH notation is an index of hydrogen's chemical activity in a solution. Basically, pH is a measure of how acidic or basic a substance is. The pH scale is a log scale and is measured from 0 to 14; 7 is neutral. The more acidic a substance is, the lower the pH it will have; therefore, substances with a pH lower than 7 are acidic. The more basic a substance is, the higher the pH. Basic substances have a pH higher than 7. As well, each whole number below 7 is 10 times more acidic than the next higher number. For example, a pH of 3 is 10 times more acidic than a pH of 4, 100 times more acidic than a pH of 5, and 1,000 times more acidic than a pH of 6. The same concept is true for basic substances, those with a pH above 7. Each whole number above 7 is 10 times more basic than the next lower number. Pure water has a pH of 7 and is considered neutral.

14. The pH of a solid can be measured only if the solid is dissolved in deionized water. That is, the pH measurement can be taken only if the substance is in liquid form.

15. Infrared imaging is one of the most versatile and widely used methods for detecting surface temperature variances caused by abnormal or uncharacteristic conditions. Infrared imaging can be used to discover leaks in pipes and other sources of energy and material loss.

16. Irreversible monitoring labels have the ability to maintain the highest level of surface temperature for future reference.

Chapter 5—Nondestructive Testing and Evaluation

1. Nondestructive testing (NDT) involves methods of testing and evaluating material properties in order to detect defects in engineering

structures without impairing their future use or altering the integrity of the material. Some of the testing techniques used in PDM are ultrasonic, thermography, radiography, Eddy current, liquid penetrant, magnetic particle, and holography.

2. In the case study, the nuclear power plants in Japan used the technique of Eddy current testing in their predictive maintenance process.

3. X rays are in the radiography category of NDT. Some of there industrial uses in PDM include evaluating the condition of forgings, welds, castings, and many other metallic and nonmetallic fabricated parts.

4. The NDT technique that "oil-and-whiting" refers to is liquid penetrant. This method works by first cleaning the test object to remove any oil, dirt, or other material that would interfere with the action of the penetrant. The liquid penetrant is then applied to the test object. The liquid penetrant usually contains some sort of dye. The penetrant is then given sufficient time to permeate the surface cracks and openings. The object is then wiped clean and a developer is added to the surface. The developer draws the penetrant to the surface, which makes the cracks visible.

5. Magnetic particle testing is used in the detection of cracks in the surface of objects by studying the imaginary lines of force formed around a magnetized steel bar. A crack in the surface of the object will disturb the lines of force and each side of the crack will behave as separate magnets, forming opposite magnetic poles on each side of the crack. (See Figures 5–5 and 5–6.)

6. No, aluminum would not be recommended for magnetic particle testing because aluminum is not magnetic.

Chapter 6—Implementing TPM

1. TPM is a corporate-wide equipment and resource management strategy for overall quality and productivity improvement. Its goal is to increase overall productivity.

2. Some fundamentals of TPM are to avoid the crisis in the first place and to shift the ownership and responsibility for the equipment and its upkeep to the operator.

3. They both include ideas of empowerment and improvement for the company as a whole. They both also stress the importance of training and education.

4. Yes, TPM can be implemented in a union environment; this can be seen by examples at AT&T, DuPont, and Ford Motor Company, to name a few examples.

5. **a.** Answers will vary.
   b. Answers will vary.

6. Answers will vary.

7. OEE is overall equipment effectiveness. The factors affecting it are availability, efficiency, and quality.

8. Cleaning is the most important step in equipment maintenance because it provides an understanding of the machine, it plays a role in identifying contaminants and their sources, and it keeps the equipment in like-new condition.

9. The purpose and importance of developing lubrication standards is that the standards ensure that lubrication is performed on a regular basis and they reduce the possibility of error or confusion.

10. The role that continuous process improvement (CPI) plays in TPM is to reduce setup and manufacturing cycle times, reduce and simplify material handling and inspection, and eliminate and minimize work in process.

11. Workplace organization means that there should be an appropriate place for every part, tool, and so on, and every item must be placed in the appropriate location. Workplace organization plays an important role in quality and productivity.

Chapter 7—TPM Implementation and Process Improvement Tools

1. Benchmarking: To search for the best in class and learn from them.

2. The basic blueprint set forth by AT&T for benchmarking is as follows:
   a. Conception: Which areas would benefit from benchmarking?
   b. Plan: Establish goals and objectives of benchmarking and create a plan.
   c. Collect data: Collect data on current process and practices.
   d. Identify best in class: Who is best?
   e. Collect data on best practices: How do these companies achieve their status?
   f. Compare: Compare your practices and processes with those of the best.
   g. Implement: Plan to incorporate their practices into your own operations and procedures.
   h. Evaluate: Assess your success.

3. Benchmarking helps a company improve their methods and practices by allowing the company to compare their procedures to those of the best and learn from the best, implementing how they do things in their own procedures.

4. I disagree because you can learn from the best and then expand on their practices, which may put you over the top, making you the best.

5. The pillars of TPM are planned maintenance, maintenance free, individual *kaizen*, education and training, and self-maintenance.

6. The eight pillars of maintenance are support and guidance for operator self-maintenance activities, reduction of breakdowns, establishment of planned maintenance activities, lubrication management, spare parts management activities, maintenance cost management activities, maintenance efficiency improvement, and training.

7. The purpose of FMEA is to identify potential failures, that is, all the ways in which equipment, a process, or a product can fail, attempting to determine how these failures can or may affect the overall operation of the equipment or the entire system, and to propose solutions or a course of action to eliminate these potential failures. The role that FMEA can play in TPM is that it trains and educates the operators on how to handle potential problems that they encounter while on the job effectively.

8. The word "potential" should be added, to make it potential failure mode and effect analysis.

9. Some potential failure modes affecting a calculator are that it will not perform math functions correctly, buttons could become jammed, and so forth. Effects that these failures could cause might include failing an exam because your calculator malfunctioned, you get the wrong answers, and so on.

10. **a.** 84 or 42
    b. 162 or 180
    c. 25

11. b, a, c

12. Root cause analysis is a systematic approach in identifying the basic or the root cause of a problem or an undesirable condition so that actions may be taken to eliminate the cause and prevent the occurrence of the undesirable event. Root cause analysis can be related to maintenance in the sense that it aims to identify the causes of potential failures and prevent them in the process.

13. Compare your answer with the text material.

14. Pareto analysis generalizes root cause analysis by stating that in most cases, 80% of the downtime is caused by failures in only 20% of the equipment.

15. The change analysis technique for root cause analysis is based on the assumption that a change in the process, material, procedure, and so on must have occurred in order to cause the problem at hand.

16. Design of experiments (DOE) is a technique that isolates a series of variables that the experimenter might suspect to be influential in a given process, then systematically manipulates these variables and their settings to determine what variables and at what settings would be optimal for the process. The relationship between DOE and maintenance functions is that the DOE will determine the root cause analysis and in turn determine what maintenance functions need to be performed.

17. Event and causal factors analysis is a systematic approach to examine events, their sequences, the conditions under which these events occurred, and any factor that might have resulted or caused that event.

18. An Ishikawa diagram can help with root cause analysis because it can be used to represent the relationship between an event and its causal factors and it helps in identifying all the possible causes for a problem.

19. "OR" gates are used to indicate that any one of the factors or causes that follow is a sufficient cause and by itself can produce the effect. "AND" gates are used to indicate that all the causes or factors that follow must be present in order for the event to occur. The difference between the two is that only one factor needs to occur for an "OR" gate and all factors need to occur for an "AND" gate.

20. A fault tree is a diagram that allows you to see all the causes, their relationships, and the necessary conditions that can result in a failure. Fault trees are used as a troubleshooting tool to gain insight into an existing problem.

21. Answers will vary.

22. This means that everyone needs to know and understand why and how the problem occurred and the necessary measures needed to correct the problem or they may be the reason that the problem occurs again.

23. Alternative solutions may be evaluated on the basis of implementation costs, feasibility, return on investment, ease and difficulty of implementation, constraints and limitations, and the probability

that the given solution will be successful in removing the hazard or eliminating its future occurrence.

24. By destroying the products, they are able to learn what happens in different situations and are able to find solutions to these problems before their products are even sold to the public, which improves the safety of the product.

Chapter 8—Facility Maintenance Projects Planning and Control

1. General operating procedures can guide the completion of repetitive tasks. Projects are generally special activities that are not performed the same way twice.

2. • They require a group of tasks to be performed concurrently and/or consecutively.
   • They can be defined by the dollar value of the activities performed.
   • They are defined by the level of effort required by multiple departments in an organization.
   • They require more planning and design effort than normal maintenance and preventive maintenance activities.
   • Resources are allocated among tasks to accomplish the project within time and budget constraints.
   • They could be delayed to a future date.

3. The project could have been delayed to a future date, as it was an optional activity pursued to reduce the long-term transportation costs for the mining company. Projects are generally improvements to cost or profit possibilities.

4. 1. What are the project objectives?
   2. What is the process to be used to complete the project?
   3. What will the result of the project be?

5. 1. Clearly define the project scope, specifications, and deliverables.
   2. Identify all activities for the project.
   3. Identify any precedence relationships for tasks.
   4. Identify resources available and assign personnel to the project.
   5. Develop time estimates for each activity.
   6. Develop the project schedule, using Gantt charts.
   7. Identify the longest path of the project.
   8. Identify any slack for the project.
   9. Manage the project.

6. The tasks that must be completed before another task can start. This is called identifying a precedence relationship.

7. Answers will vary.

8. The longest path of the project. Add up each path and the longest path becomes the critical path, in that its completion determines the total length of the project. This path will determine whether or not the project can be completed on time. Any slippages along the critical path might result in the project being late, depending on how much slack there is between the total length of the critical path and the time allotted to complete the project.

9. Internal boundaries include any exchange made between the project team members and the project manager. External boundaries are any communication between the project team or project manager and customers or any stakeholders of the project.

10. Work items that must be completed in the upcoming period and long-term work items, which are part of a larger project effort.

11. Answers will vary.

12. The critical path method dates back to 1957, when it was developed by Remington Rand and DuPont for use in the construction and maintenance of DuPont chemical plants. The program evaluation and review technique was developed in 1958 by the Navy for planning and control of the Polaris missile submarine program.

    Steps in the conduct of both the CPM and the PERT network analyses are similar. The differences in the two approaches involve the estimated times to complete each task, or "activity" of a project. CPM is deterministic, assuming fixed-time estimates for the activities are sufficiently accurate. PERT is stochastic (probabilistic), eliciting probability distributions for the task times.

13. Crashing an activity means using extra resources to reduce the duration of an activity.

CPM Crash Analysis

14 and 15.

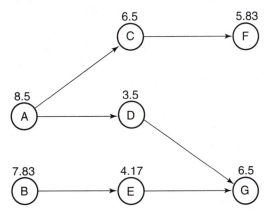

a. Project time = 20 weeks; critical path is A → C → F; cost is $35,700.

b. Step 1, decrease F to 4 weeks; cost is now $35,700 + 800 = $36,500.

Step 2, decrease A to 8 weeks and E to 4 weeks; cost is now $36,500 + 1,800 = $38,300. *T* is now 18 weeks, and *all* paths are critical: A → C → F and A → D → G and B → E → G

16.

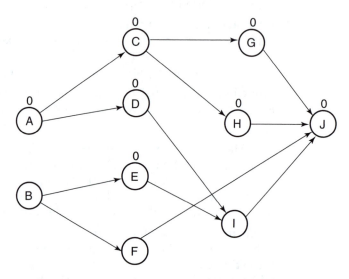

a. Normal project time is 43 days, cost of $9,600 along path

A → C → H → J.

b. Full-crash *T* = 36 days, cost is $13,075 along *two* critical paths:

A → C → H → J and A → D → I → J

c. Minimum project time of 36 days can be achieved at a cost of $12,100 by reducing A from 10 to 8, C from 13 to 10, E from 12 to 11, H from 7 to 6, I from 8 to 5, and J from 13 to 12. So B remains at 8, D remains at 11, F at 15, G at 15. There are *three* critical paths for this minimum-cost/ minimum time solution:

A → C → H → J and A → D → I → J and B → E → I → J

d. Min Time/Min Cost savings over full crash is $ 13,075 − $ 12,100 = $ 975.

PERT Analyses

17. With mean $T = 118$ weeks and standard deviation of 12 weeks:
    a. $P(T > 136) = .0668$, or about 7% (area enclosed *within* $Z = 1.5$ is .4332).
    b. $P(T <= 114) = 37\%$ (area *within* $Z = -.333$ is interpolated to about .1304).

18.

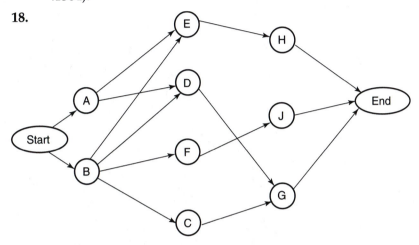

a. Mean T is 21 months along critical path $B \rightarrow E \rightarrow H$ (an activity-on-arrow network diagram would have a dummy activity arrow from the end of B to the beginning of D and E).
    b. Project time standard deviation is 2.19 months (rounded from 2.1858, which is the square root of the project variance, calculated as 172/36, or 4.7777).
    c. $P(T > 25) = .0336$, or about 3%. So the project manager should not be worried.

    (The Z value for 25 is 4/2.19, or about 1.83, which encloses .4664.)

19. Network calculations give a mean project time of 44 weeks with a variance of $214/36 = 5.9444$ weeks. Thus, standard deviation is 2.45 weeks.
    a. $P(T <= 42) = .2061$ (Z value for 42 is about $-.82$, enclosing .2939 area).
    b. $P(T > 48) = .0505$ (Z value for 48 is $+1.633$, enclosing about .4495 area).

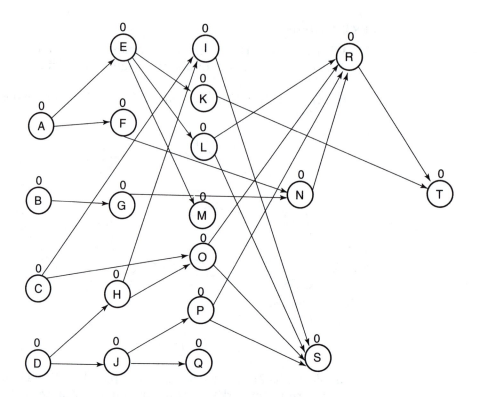

Chapter 9—Computerized Maintenance Management Systems

1. CMMS are used for all aspects of maintenance planning, management, and control, including maintaining equipment history, scheduling preventive and predictive maintenance activities, spare parts inventory management, and a variety of other maintenance activities.

2. With inventory management and tracking, it can help keep track of part numbers, stock location, quantities on hand, and unit cost.

3. A CMMS package can maintain a record of craft skills, training, and education information such as course description hours and credit units earned and other relevant data for each individual.

4. CMMS can be useful because all information for a given piece of equipment can be put into the system, including maintenance dates, warranty information, service contract records, and failure dates.

5. All safety information on equipment can be stored in CMMS and all related OSHA regulations that affect the organization are

maintained and included in the work order as reminders of the regulatory requirements.

6. It allows scheduling to be based on equipment needs, availability of personnel, and production schedule. It also allows for specifying workdays that are not available due to employee vacations and other variables.

7. With CMMS, work orders are generated automatically for scheduled preventive maintenance actions or may be created manually for corrective and emergency situations.

8. The statistical capability of most CMMS packages is that they are able to identify equipment readings out of the control limits and alert you to schedule the required maintenance before failure occurs, they can produce graphs and reports showings the readings and trends of a specific metric, and they generate work orders.

9. ISO standards now require the use of CMMS to become certified. With CMMS, companies can document standard operating procedures and adhere to them, which is a requirement for ISO certification.

INDEX